W9-AQC-288

Power, Speed, and Form

Copyright © 2006 by Princeton University Press

Published by Princeton University Press, 41 William Street, Princeton, New Jersey 08540

In the United Kingdom: Princeton University Press, 3 Market Place, Woodstock, Oxfordshire OX20 1SY

All Rights Reserved

Library of Congress Cataloging-in-Publication Data

Billington, David P.

Power, speed, and form : engineers and the making of the twentieth century /

David P. Billington and David P. Billington Jr.

p.   cm.

Includes bibliographical references and index.

ISBN-13: 978-0-691-10292-4 (hardcover : alk. paper)

ISBN-10: 0-691-10292-9 (hardcover : alk. paper)

1. Engineering—United States—History.   2. Engineers—United States.

I. Billington, David P., 1953– .   II. Title.

TA23.B48 2006

620.00973—dc22

2005037085

British Library Cataloging-in-Publication Data is available

This book has been composed in Electra, Myriad and Coronet

Printed on acid-free paper. ∞

pup.princeton.edu

Printed in the United States of America

3   5   7   9   10   8   6   4

Power, Speed, and Form

Engineers and the Making of the Twentieth Century

DAVID P. BILLINGTON AND DAVID P. BILLINGTON JR.

PRINCETON UNIVERSITY PRESS
PRINCETON & OXFORD

To Phyllis, Gerda and Jonathan, Jane and Nelson
who bring the past into the present

Contents

List of Sidebars

List of Figures

Preface

To most Americans, the word *technology* has come to mean computers and the electronic networks associated with them. A look at the top five American companies in the Fortune 500 of 2004 may therefore come as a surprise: four of them (General Motors, Ford Motor Company, ExxonMobil, and General Electric) were the leading firms seventy-five years ago, and none (the other is Wal-Mart) is primarily engaged in the business of making or networking computer hardware and software.[1] The leading technologies three-quarters of a century ago are still mainstays of the American economy today. This book describes the innovations from 1876 to 1939 that launched these technologies: the electric light and power network, the telephone, oil refining, the automobile, the airplane, radio, large steel bridges, and reinforced concrete. The book is a sequel to *The Innovators*, which covered American engineering from 1776 to 1883.[2] The two books explain the principal engineering ideas that helped transform the United States from an agrarian society in the eighteenth century to the industrial civilization it became in the twentieth.

The word technology also connotes objects that people can use, while the word *engineering* is more remote and intimidating. This remoteness reflects the isolation of the engineering profession from society. Technological literacy is now a goal in many universities and colleges, yet all too often this literacy is left to distributional courses in science and mathematics that contain little or no engineering, and some schools think of technical literacy merely as an ability to use personal computers. Few schools recognize a need to give all students a basic exposure to modern engineering, and if more educators did see such a need, few engineers would know how to help them meet it. Introductory science courses for non-scientists can be found in every college and university, but introductory engineering courses for non-engineers are rare. Traditions in engineering education and in broader academic life perpetuate the isolation of engineering on most campuses that teach it, and the efforts of institutions to promote technological literacy still leave the essence of engineering inaccessible to a broader public.

As part of an effort to remedy these problems, this book highlights what we believe to be key engineering ideas and events in the years between 1876 and 1939. Our goal is to explain to a non-technical audience, and to engineers themselves, the ideas behind historic innovations that are still essential to modern life. We explain these ideas in the language of engineering—that is, mathematical formulas—but unlike engineering textbooks the formulas we employ do not require calculus. Some of the innovations we describe will be familiar and some will not be well known. As far as we know, these innovations have never been collected together in a one-volume overview that presents them numerically as well as narratively. Our book should be useful to engineering faculty who need a text for an introductory course in engineering, and it should be especially useful as a text for non-technical students who take such a course to fulfill a distributional requirement in science and technology. We also hope that our book will be helpful to high school science teachers who would be interested in teaching engineering.

Our book is not a comprehensive history of the engineering in our period. In *They Made America*, Harold Evans provides outstanding accounts of engineers whose work transformed American life in the nineteenth and twentieth centuries.[3] A new undergraduate textbook, *Inventing America*, also weaves invention and inventiveness into the larger tapestry of American history.[4] We examine engineering and its innovators with a different focus. Our aim is to provide a numerical as well as narrative grounding in technical ideas that are basic to modern civilization and in this way to give more of the engineering story of the people who conceived them. We believe that our approach introduces engineering in an engaging way and that it will help efforts to attract new students to engineering careers.

Our book employs a framework of terms with which readers of *The Innovators* will be familiar: the four ideas of structure, machine, network, and process that we use to characterize the principal works of modern engineering. This framework makes engineering more comprehensible, and we recapitulate it in our first chapter. In this volume, however, there is a more fundamental theme that needs to be underlined in a preface: the distinction between engineering and science.

Science is *discovery*, the study of what already exists in nature. Engineering is *design*, the creation of objects and systems that do not occur naturally. Scientists in the early nineteenth century discovered laws of electricity and electromagnetism that enabled Thomas Edison to invent a new electric lighting system from 1878 to 1882. But these discoveries did not tell Edison how to design his system. In fact, Edison's ideas of a high-resistance bulb and a parallel circuit challenged the views of influential scientists and engineers in the 1870s. We show in nearly all of the cases we describe that the contributions of science to new technologies were more significant *after* the original breakthroughs than immediately before them. Where preceding scientific advances were important, the science usually preceded the innovations by several decades. Scientists and engineers both need to understand the natural world and both use mathematics. But science and engineering have different purposes and innovation is not simply the application of basic science.[5]

In treating engineering as design, we follow the engineer and historian Walter Vincenti in making a further distinction between *normal* and *radical* design.[6] The former refers to incremental improvements to an established technology, whereas the latter consists of the less frequent ground-breaking innovations that create or establish new technologies. The distinction is not absolute, and radical innovations never occur in a vacuum. But the distinction is still useful: the Stearns duplex telegraph of 1872, which enabled two messages to travel over a single telegraph line in opposite directions at the same time, was what we would call a normal innovation, because it was primarily an improvement to an existing technology, the telegraph. Alexander Graham Bell's 1876 telephone, which used a wire line to transmit voice using different kinds of sending and receiving equipment, was what we would call radical. Our focus is on radical ideas.

Most of the radical innovations we cover have been the stuff of engineering history for a long time. But a crucial aspect of them has never received the emphasis it deserves: the simplicity of the basic ideas. In examining their work, we found that the engineering innovators of our period described their work in terms of surprisingly simple formulas or concepts. There is good reason why they did so: with innovations so new and unfamiliar, the engineers were concerned with expressing them as accessibly as possible. Later engineers, often with the help of scientists, made the new technologies more sophisticated and efficient. But the innovations began in most cases with insights at the level of secondary school mathematics.

This clarity is vital to bring out for two reasons. First, it tells us that simplicity, not complexity, is the characteristic of original engineering thought. Second, through examples of such thinking, students can learn ground-breaking engineering ideas without first having to know calculus and physics. No one can advance to a professional level in modern engineering without meeting its demands for knowledge of more-advanced science and mathematics. But at the entry level, engineering in the United States has turned away the new people on whom its future depends by an abstract and narrow approach to introductory teaching. Students can be more attracted to engineering when their first exposure to it is through historic examples that are still relevant to modern life. Such examples integrate different branches of engineering and convey how engineers think. Students who do not intend to become engineers receive a real literacy in modern engineering when they too see and work with original concepts and the numbers that express them.[7]

Modern engineering is more coherent and more appealing when it is introduced as a historical sequence of ideas and events. Engineering education follows the model of the physical sciences, in which detailed knowledge proceeds in an abstract sequence from general principles. We believe that modern engineering is better introduced as a narrative of great works, like the history of art. This narrative points to a tradition. The law, medicine, art, architecture, politics, and the natural sciences all have traditions of canonical ideas and events. In contrast, all too many engineers think of their knowledge as a body of information perpetually existing in the present tense. Too many do not realize that there is a canon, a grand tradition, in modern engineering consisting of great ideas that can be expressed in accessible formulas. Engineers and the general public need to learn these ideas no less than they need to learn the great ideas and events that are the heritage of the natural sciences, the social sciences, and the humanities. This tradition also points to the crucial role of individuals. Modern engineering would not have been possible without the contributions of larger groups of people, and interactions with society and culture have shaped it in crucial ways. But creative individuals remain vital to the tradition of modern engineering, and students need to know that they can learn from and emulate the best engineers.

The popular image of technology is one of accelerating change. Yet great technical ideas of the past often renew themselves in new forms. The reciprocating engine of the automobile uses the same principle as the reciprocating steam engine that pulled railroad trains. Although more sophisticated and more powerful than the nineteenth-century telegraph network, the Internet of today is also at its foundation a network of wired and wireless circuits that transmit information. The image of accelerating change neglects these deeper continuities. But a technological society is not a predetermined one. Unlike a mathematics or physics problem that has only one right answer, an engineering formula does not define a "one best way." Numbers and natural laws define limits; but in every technical problem, there is room for choice between alternatives that make engineering sense. With this choice comes freedom and also responsibility. Too often engineers neglect the humanistic impact of their work, just as critics of technology too often dismiss its humanistic potential. Our book is part of an effort to connect the two cultures.

Acknowledgments

This book began as a series of lecture notes for a course begun by the senior author in 1985, now entitled "Engineering in the Modern World," that is taught at Princeton University every year, primarily to first-year undergraduate engineering students and liberal arts undergraduates. The course originated in the "New Liberal Arts" program begun by the Alfred P. Sloan Foundation in the early 1980s under its president Albert Rees. Financial support from the Sloan Foundation began and sustained the research and writing that led to this book. For Sloan support in recent years, the authors are indebted to Doron Weber. They are also grateful to the National Science Foundation's Division of Undergraduate Education, and to Norman Fortenberry and William Wulf of the National Academy of Engineering, for support of the teaching and research on which this book is based.

"Engineering in the Modern World" was fortunate in its early years to have the backing of colleagues of the senior author, including Bradley Dickinson and Paul Prucnel in electrical engineering; John Gillham, Roy Jackson, and Richard Golden in chemical engineering, and Frediano Bracco and H. C. Curtis in mechanical and aerospace engineering. Three non-technical colleagues, Hal McCulloch, Peter Bogucki, and Tom Roddenbery, joined later as preceptors and contributed a liberal arts perspective that has helped make the content more accessible to all students. As the present volume neared completion, Roland Heck, a chemical engineer, began to teach in the course and made a vital contribution to our chapter on petroleum refining.

Introductory engineering courses have been difficult to sustain in most schools, mainly because they are conceived and conducted as experimental rather than permanent courses. To become permanent, they must find a place in the core curriculum of their institution, to fulfill either an engineering requirement or an undergraduate requirement in science and technology. The support of Princeton president Harold

Shapiro and the faculty Council on Science and Technology, led first by the late David Wilkinson and then by Shirley Tilghman, now president of the University, proved critical to achieving this recognition for "Engineering in the Modern World" and to giving the course its unique foundation in research and strength in teaching. The course is now a way for undergraduates to meet the university laboratory requirement in science or technology, and as a result it is one of the most heavily enrolled at Princeton. Deans James Wei and Maria Klawe of the School of Engineering and Applied Science have helped the course and its associated scholarship make the contribution we intend to the goal of transforming the nation's engineering education and attracting and graduating new kinds of engineers from Princeton and other engineering schools.

Since 1996 the senior author has co-taught the course on a regular basis with Michael Littman, a colleague in mechanical and aerospace engineering at Princeton, whose innovation of teaching laboratories for the course was pivotal to its recognition as a science and technology laboratory course. In addition to stimulating us to think about new and engaging ways to explain engineering, Mike has patiently reviewed successive versions of the manuscript for this book and has greatly improved its accuracy in the fields least familiar to the authors. The two authors have benefited as well from a number of outside scholars. Terry Reynolds of Michigan Technological University kindly read an earlier draft of this book and offered very helpful comments. Paul Israel, editor of the Thomas Edison Papers at Rutgers University, read two of our chapters twice and gave vital insight into the work of Edison, Bell, and other electrical engineers. David Wunsch of the University of Massachusetts at Lowell reviewed two versions of the chapter on radio and gave invaluable knowledge and advice, and William Case of Grinnell College also provided vital assistance on our radio chapter.

Philip Felton gave essential help understanding the modern automobile engine, and H. C. "Pat" Curtiss shared his experience and insight in aerodynamics and the work of early aviators and aeronautical engineers. The chapter on Othmar Ammann and the George Washington bridge grows out of an article written by the senior author with Jameson Doig, former chair of the Politics Department at Princeton, that received the Usher Prize from the Society for the History of Technology. Dr. Margot Ammann Durer generously provided knowledge of her father. Donald C. Jackson of Lafayette College brought the extraordinary work of John Eastwood to our attention and also gave this book a valued review, and the material on Tedesko was greatly helped by joint research with Eric Hines and with Edmond Saliklis of the California Polytechnic State University at San Luis Obispo. Although these individuals have been generous in giving us their assistance, any errors and faults that remain are our own and not theirs.

Former colleagues at Princeton whose support for the senior author's teaching made this book possible include Norman J. Sollenberger, Joseph Elgin, Robert Mark, John Abel, Ahmet Cakmak, and Peter Jaffe. Colleagues at other schools whose work has contributed include John Truxal, Marian Visich Jr., Alfonso Albano, J. Nicholas Burnett, Newton Copp, Andrew Zanella, and Atle Gjelsvik. The historians Carl Condit, Edwin T. Layton Jr., Merritt Roe Smith, Robert Vogel, and George Wise have also provided valuable assistance. We would also like to thank Professor Andrew Wood of San Jose State University for his help, and King Harris and Ann Richardson for information about their father, King Harris, and uncle, Lawrence Harris.

The senior and junior author are deeply grateful to our editor, Sam Elworthy, and to Deborah Tegarden, Pamela Schnitter, Dmitri Karetnikov, Brian MacDonald, Alycia Somers, Shani Berezin, and Maria denBoer for their assistance and support in bringing this book to publication.

The book could not have been produced without the teaching assistants who have made "Engineering in the Modern World" so successful at Princeton: Scott Hunter, Christopher Peck, Ronald Wakefield, Rosemary Secoda, John Matteo, Roger Haight, Karen Mielich, Susan Lyons, Nicholas Edwards, Daniel Chung, James Guest, Michael Tantala, Gayle Katzman, Nicolas Janberg, Moira Treacy, John Ochsendorf, Chelsea Honigmann, Ryan Woodward, Richard Ellis, Maria Janaro, David Wagner, Sinead Mac Namara, Gregory Hasbrouck, Nicole Leo, Kristi Miro, Powell Draper, Shawn Woodruff, Michael Bauer, Sarah Halsey, Rebecca Jones, David O'Connell, and Allison Schultz. The research and teaching of the course has also had the benefit of undergraduate research and archival assistants Michael Starc, Abbie Liel, Angela Ovecka, William Cooch, Taylor Greason, Jennifer Bennett, Eve Glazer, and Diana Zakem. Joseph Stencel and Joseph Vocaturo have been indispensable as course administrator and laboratory director respectively. Powell Draper, Sinead Mac Namara, Kristi Miro, Shawn Woodruff, and Jennifer Bennett provided special research assistance for this book.

"Engineering in the Modern World" owes a special debt to J. Wayman Williams for research and for the design of outstanding instructional materials. Kathy Posnett has given vital administrative support, and the librarians and library staff of Princeton University deserve special thanks for accommodating the needs of the course and the library research for this book. To the senior author's brother, James H. Billington, the thirteenth Librarian of Congress, both authors are deeply indebted for essential counsel, guidance, and support.

After completing his doctorate in modern history from the University of Texas at Austin, the junior author accepted the senior author's invitation to rewrite a first draft of this book. Through further research and rewriting, the junior author discovered that original engineering ideas could be grasped by someone with a liberal arts education and that history is central to an integrative understanding of modern engineering. The book has been a collaborative effort not only between the two authors but between them and the colleagues, students, and anonymous peer reviewers without whose crucial assistance the finished work would not have been possible.

The junior author's scholarly vocation owes a special debt to the guidance of Professors Cyril Black and Julian Boyd of Princeton University and to Stephen C. Flanders of WCBS Radio in New York and Roswell B. Wing of the U.S. Department of Commerce in Washington, DC. The junior author is also deeply grateful to Neva Wing, Carol Flanders, Corinne Black, and Mary Laity for their encouragement and support.

The New York State Board of Regents External Degree Program (now Excelsior College) gave the junior author the chance to finish an undergraduate degree by examination during a convalescence. At the urging of Jay Bleiman of the Woodrow Wilson School at Princeton, and with the help of S. Frederick Starr, the junior author attended the Paul H. Nitze School of Advanced International Studies in Washington, DC. For a synoptic grounding in world affairs, he owes particular thanks to Bruce Parrott and Michael Vlahos, and to Michael Harrison, Kendall Myers, Charles Pearson, Stephen Szabo, and Nathaniel Bowman Thayer. To Marc Rothenberg and the staff of the Joseph Henry Papers at the Smithsonian Institution; and to the staff of the Woodrow Wilson International Center for Scholars, particularly Zdenek David, Prosser Gifford, Ann Sheffield Roth, and George Liston Seay, he owes the unique opportunities to participate in scholarly work that motivated him to pursue advanced study in history. At the University of Texas at Austin, he owes an immense debt to

W. Roger Louis, his adviser, and to Standish Meacham and the other members of his committee William Braisted, Robert Divine, and Robert Hardgrave, and to Nancy Barker, David Crew, Bruce Hunt, John Lamphear, Brian Levack, Howard Miller, Sidney Monas, Walt Rostow, Claudio Segré, and Philip White. The junior author could not have succeeded as a scholar and student without the generous friendship and support at U.T.-Austin of the many graduate students and spouses he came to know, who will be thanked properly in the published version of his doctoral dissertation. To Travis Hanes, Sharon Arnoult and Billy Branch, Peter and Lauren Austin, Aaron and Takako Forsberg, and Jennifer Loehlin he owes special thanks for their continuing interest.

The junior author is deeply grateful for the friendship and support over the years of the Flanders family—Stephen and Hedy, Jeff and Maisie, Tony and Bunny, Julie and Emil Adler, and Carl and Andrea; the Laity family—Jim and Mary Ann, Susan, Kate, Bill, and John; the Black family—Jim and Martha, and Christina; and of Orest Pelech, Tom Ruth, Glenn Speer, and Aaron Trehub.

For their love and support through many stages of life, the junior author owes a very deep and personal debt to Marjorie Billington, John and Lynn Billington, Janet and John Fisher, and Arloa Bergquist, and to all of his Billington and Bergquist cousins.

The junior author's sisters and brothers Elizabeth and Donald, Jane and Johnson, Philip and Ninik, Stephen and Miriam, and Sarah and Peter, and most especially his mother Phyllis, through their love helped sustain the years of effort that went directly or indirectly into this book. The junior author's hope is that this book will join the contributions of faculty and scholars at other institutions in a continuing and widening effort to help improve the education of engineers and liberal arts students, and to bridge the divide in understanding that separates engineering from the society it serves.

CHAPTER ONE

The World's Fairs of 1876 and 1939

O n a much-anticipated trip in 1939, a family of four drove from the Philadelphia suburbs up U.S. Route 1 into New Jersey and then turned east through the dark overpasses of Weehawken. Suddenly the road began a sweeping circle to the right, and over the left parapet, lit by the afternoon summer sun, appeared the skyline of Manhattan with the towers of the Empire State and Chrysler buildings and Rockefeller Center. Soon the car descended into the three-year-old Lincoln Tunnel before emerging in New York City itself. The senior author (age twelve) and his brother (age ten) had their first view of New York City on the way to Flushing Meadow, site of the 1939 New York World's Fair. But first they checked into the Victoria Hotel in Manhattan where they were transferred to somewhere in the sky—probably the thirtieth floor—with a view they had never before experienced. They were in the skyline.

The next morning the family drove to the fair and went to the most popular exhibit, the Futurama ride in the General Motors pavilion, also named Highways and Horizons.[1] Although they were early, there was already a line longer than either preteen could see. But they were fresh, and finally their turn came. Into the cushioned seats they nestled, and a smooth voice guided them through an incredible landscape of highways, skyscrapers, parks, cities, factories, and forests, all in miniature and all seen from above as if in a low flying airplane (figures 1.1 and 1.2). The unforgettable sixteen minutes in the grip of Norman Bel Geddes, the industrial designer, surpassed all of the other impressions of the fair. It was the future. Even in youth one could sense the central themes of mobility, speed, and adventure—an urban frontier of new cities and new sceneries. Children of the Depression, the senior author and his brother had traveled very little and had led simple lives; their parents would often point to the impressive stone building in Narberth, Pennsylvania, where they had lost all their savings in the bank crisis. The bank lobby was now a beauty salon.

Figure 1.1. James and David Billington at the 1939 New York World's Fair. Source: Billington family album.

Figure 1.2. The Futurama ride at the 1939 General Motors pavilion. Courtesy of Professor Andrew Wood, San Jose State University, and General Motors.

Those who remember the Futurama ride have lived to see it realized as they circle above any major airport today. There below are the superhighways, the tiny cars, the skyscraper cities, and the vast suburbs. In 1939 this civilization could be glimpsed; a generation later it was reality. Yet the 1939 New York World's Fair was not just a vision of the future; it portrayed a technology and a society already in existence. The cars, highways, buildings, and industries that Bel Geddes portrayed were familiar objects. The Futurama ride was a celebration of steel and concrete, oil and cars, flight and radio, and above all electric power and light—the great industrial innovations of the late nineteenth and early twentieth centuries.

Before these transformative changes, America was a mostly rural society. People lived close to nature and prosperity depended on the harvest. Technology was simple: most houses were built out of wood or stone, firewood supplied fuel, candles gave light, and tools and equipment were made of wood and iron. Local transport was by horse or horse-drawn carriage over unpaved roads. Life was slow-paced, and communication— for those who could read and write—was by letter. But the steamboat and then the railroad and the telegraph had begun to reduce isolation and to accelerate the pace of American life. An earlier world's fair, the Philadelphia Centennial Exhibition of 1876, celebrated these changes and foreshadowed even greater ones to come.

On May 10, 1876, the "United States International Exhibition" opened in Fairmount Park, Philadelphia. The fair commemorated the one hundredth anniversary of the American Revolution and came to be known as "The Centennial." Large halls dedicated to horticulture and crafts reflected a nation that was still largely rural and self-sufficient. But the principal attraction of the fair, Machinery Hall, displayed the products of new industries that were beginning to remake society. At the center of Machinery Hall stood the Corliss steam engine, thirty-nine feet high. President Ulysses S. Grant and the visiting Emperor Dom Pedro of Brazil inaugurated the fair by turning on the engine: its two giant pistons turned a huge wheel that powered other machinery in the hall. Built by George Corliss of Providence, Rhode Island, the great engine was the ultimate expression of the steam engineering that had led the first hundred years of America's industrial growth (figure 1.3).[2]

The first working steam engine was invented by Thomas Newcomen in 1712. Steam from a separate boiler entered a cylinder on one side of a piston. Applying cold water condensed the steam and created a partial vacuum. Atmospheric pressure on the other side then pushed the piston and pulled a rocking beam, enabling the engine to pump water from mines. In 1769 James Watt created a separate condenser that allowed the temperature of the cylinder to remain relatively constant. Watt's engine could perform the work of the best Newcomen engines with about one-third of the fuel.

Watt soon designed a version of his engine to turn wheels, providing rotary motion to run factories. Belts connected to the wheels of steam engines soon began turning grain mills, weaving looms, machine tools, and other equipment. Early factories in America did not at first need steam power. Water from nearby rivers gave Francis Lowell and other New England manufacturers the power they needed to create a major new textile industry. By the late nineteenth century, though, many factories had shifted to steam. The Corliss engine on display in Philadelphia was one of the largest rotary steam engines ever built.

Figure 1.3. The Corliss engine at the 1876 Philadelphia Centennial Fair. Courtesy of the Print and Picture Collection, Philadelphia Free Library. Circulating File.

The greatest contribution of steam was not to drive engines in place but to power mobile engines on boats and trains. The first great American engineering innovation, Robert Fulton's *Clermont* steamboat, proved itself in an 1807 trip up the Hudson River from New York to Albany. Steamboats soon opened the Mississippi and Ohio rivers to commerce, creating the world captured by Mark Twain, who began his own career as a steamboat pilot. Ocean going steamships soon carried much larger quantities of goods over long distances. More influential still was the railroad. George and Robert Stephenson of England built locomotives in the 1820s that used steam under high pressure to push pistons directly. During the 1830s and 1840s, railroads and steam locomotives revolutionized transportation in Britain and spread to the United States and other countries. In the 1850s J. Edgar Thomson, chief engineer and later president of the Pennsylvania Railroad, built a rail line across the Allegheny Mountains that brought the U.S. rail network to the Midwest. A transcontinental railway connected the two coasts of the United States in 1869. Locomotives designed by Matthias Baldwin of Philadelphia carried much of nineteenth-century America's rail traffic and Baldwin locomotives were prominent at the 1876 Centennial (figure 1.4).

But the 1876 fair marked the high point of the reciprocating steam engine and the beginning of its decline, for a new kind of engine made its first public appearance in Philadelphia that summer. Developed by the German engineer Niklaus Otto, the new engine also employed piston strokes. However, instead of burning fuel in a separate boiler to produce steam, Otto's engine burned fuel directly in a piston cylinder, pushing the piston head in a series of timed combustions. In the 1880s engineers began to use such internal-combustion engines to power automobiles, and in the first two decades of the twentieth century, Henry Ford made an automobile with such an engine that was rugged and cheap enough to reach a mass market. The Ford Model T and other mass-produced cars released transportation from the confines of the rail network and gave a sense of personal freedom to Americans that would define their way of life in the new century.

The Otto engine burned coal gas but internal-combustion engines soon ran on gasoline, a distillate of petroleum. Petroleum refining had grown in the 1850s to supply kerosene, another distillate of crude oil, to indoor lamps for burning as a source of light. Kerosene provided a better illuminant than candles and was more abundant than whale oil. The drilling of underground crude oil deposits in western Pennsylvania in 1859 brought a rush of small drillers and refiners to the region. John D. Rockefeller's Standard Oil Company of Cleveland, Ohio, soon dominated the industry, growing from a local refiner in 1870 into a national monopoly a decade later. Standard Oil's market for kerosene gradually declined as electric power spread and made indoor electric lights an alternative to kerosene. But the automobile would give the oil refining industry a new and even greater market in the twentieth century.

The year of the Centennial was the year Thomas Edison set up his research laboratory at Menlo Park, New Jersey. His greatest inventions were still in the future: the

Figure 1.4. Baldwin locomotive at the Centennial Fair. Courtesy of the Print and Picture Collection, Philadelphia Free Library. No. II-1342.

phonograph (1877), the carbon telephone transmitter (1877), an efficient incandescent light (1879), and the electric power network to supply it (1882). But by 1876 Edison had established a reputation as an inventor through improvements he had made to the electric telegraph. Developed in the 1830s and 1840s by Samuel F. B. Morse, the telegraph revolutionized communications in the nineteenth century. One company,

Western Union, dominated long-distance telegraphy by the 1870s. The Philadelphia Centennial fair had its own telegraph office, and telegraph devices, lines, and poles were on display. Yet the Centennial marked the high point of the telegraph too, for it was at the Philadelphia fair that Alexander Graham Bell gave his first public demonstration of a telephone. With the telegraph, messages had to be sent and received in offices by trained operators and then delivered by messengers. The telephone permitted instant two-way communication by voice. Bell's company eventually replaced Western Union as the telecommunications giant of the United States, and daily life in the twentieth century would come to depend on the telephone as much as on the car.

Like Edison, Andrew Carnegie was a telegraph operator early in his career. Carnegie rose in the 1850s from telegrapher in the Pittsburgh office of the Pennsylvania Railroad to manager of the office himself. Striking out on his own, he left the railroad in 1865 to form the Keystone Bridge Company, which built bridges across the Ohio and other rivers. Models of Keystone bridges were on display at the Centennial (figure 1.5). The enormous market for steel soon induced Carnegie to manufacture it, and he built the world's largest steel plant near Pittsburgh in 1875. He sold his firm to the New York banker J. P. Morgan in 1901, who merged it with rivals to create the first great twentieth-century corporation, United States Steel. Steel made possible the tall skyscraper buildings and long-span bridges that reshaped the cities and landscape of the twentieth century.

These breakthroughs were not the only technically significant events of the late nineteenth and early twentieth centuries. But the telephone, the electric power network, oil refining, the automobile and the airplane, radio, and new structures in steel and concrete were the innovations that set the twentieth century apart from the nineteenth. Some of these innovations and the individuals who conceived them are more familiar than others: most Americans have heard of Bell and the telephone but few will have heard of Othmar Ammann, whose George Washington Bridge connected New York to New Jersey in 1931 and became the model for large suspension bridges. This book describes these innovators and their work.

The book also explains innovations in engineering terms. Modern engineering can be grouped into four basic kinds of works: structures, machines, networks, and processes. A *structure* is an object, like a bridge or a building, that works by standing still. A

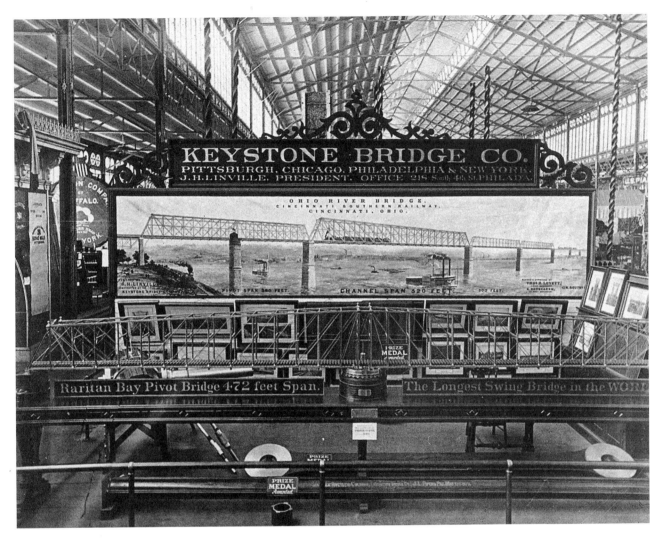

Figure 1.5. Keystone Bridge Company exhibit at the Centennial Fair. Courtesy of the Print and Picture Collection, Philadelphia Free Library. No. III-2339.

machine is an object, such as a car, that works by moving or by having parts that move. A *network* is a system that operates by transmission, in which something that begins at one end is received with minimal loss at the other end (e.g., the telephone system). Finally, a *process* operates by transmutation, in which something that enters one end is changed into something different at the other end (e.g., oil refining). We will see that many innovations are combinations of these four ideas. With structure, machine, network, and process as a basic vocabulary, complex objects and systems can be understood in terms of their essential features.

Engineers describe their works with numbers, and certain numerical relationships or formulas characterize the key engineering works of the late nineteenth and early twentieth centuries. These formulas do not explain any object or system in detail; professional engineers today would use more complex mathematics to analyze and design things. But the formulas in this book each convey the basic idea of a key work and enable the reader to think about great works of technology as engineers would think about them. In this way, the reader can enter into the imagination of the designers and can understand the basic choices that went into a design.

The Model T automobile of 1908 is an example. The car's engine had four cylinders, each containing a piston attached to a single crankshaft underneath the engine. In a timed sequence, gasoline entered each cylinder for ignition and the combustion pushed each piston downward and turned the crankshaft. Another shaft running the length of the car transmitted the rotary motion to the axles and the wheels. A simple relationship expressed by the formula $PLAN/33,000$ represents this activity and gives the *indicated horsepower* of the engine. Combustion creates a pressure P (in pounds per square inch) on the head of the piston in each cylinder. The piston travels down the length L of the cylinder (in feet). The area of the piston head A (in square inches) and the number of power strokes per minute N provide other essential information, and dividing by 33,000 gives the indicated horsepower. The formula expresses the basic working of the car's engine. In chapter 5, we give the $PLAN$ numbers for the Model T and explain how the car was efficient for the needs and conditions of the time.

A suspension bridge has no moving parts but can be explained by a formula that relates its weight and size. A suspension bridge typically has two towers that carry a roadway deck from cables anchored on each side of the bridge. The weight of the deck and of traffic on the cables tend to pull the towers inward and must be resisted by the anchorages. This resistance is called the *horizontal force*. It can be calculated in pounds by multiplying the weight on each cable q (in pounds per foot of length) by the square of the length L between the tower tops (in feet) and then dividing by eight times the depth d, the vertical distance (in feet) from the midspan of the deck to the level of the tower tops. We explain in chapter 8 how Othmar Ammann used the relationship $qL^2/8d$ in his design of the George Washington Bridge.

The *PLAN* formula did not begin with the automobile; James Watt devised it to measure the horsepower of a piston steam engine. The bridge formula also goes back to the early nineteenth century, when Thomas Telford designed the first modern iron arch and cable bridges with it. These formulas show that earlier engineering ideas often find new forms and uses, but never lose their relevance. The engineering ideas we present from the past are still fundamental to technology today.

But numerical relationships are not just matters of calculation. They involve choices. A car can be designed to be expensive or affordable, rugged or stylish. A bridge can hold up its weight with a design that is costly or economical, ugly or elegant. The best engineers strive for more than just efficiency. They value economy in the cost of making and operating an object or system, and when the needs of efficiency and economy are met, they look if possible for elegance.

Many of the engineers in our book were opposed by experts immersed in established technologies and ways of doing things. Thomas Edison challenged engineering authorities who did not think his system of light and power was scientifically possible. Unlike professional telegraph engineers, who believed that what society needed were more efficient forms of telegraphy, Alexander Graham Bell saw the potential of a telephone. William Burton faced the opposition of the Standard Oil directors, who saw no need for a more efficient way to obtain gasoline from oil, and Henry Ford had to overcome investors who believed that reaping profits from the small market for luxury cars made more sense than producing an affordable car to serve a mass market. The Smithsonian Institution for many years denied the Wright brothers credit for inventing the airplane, and in his efforts to combine safety and elegance in dam design, John Eastwood had to struggle against a narrow engineering opinion that looked for safety in massive works.

In some cases, though, the innovators were the problem. Edison clung to his system of electric power distribution using direct current, giving George Westinghouse the opportunity to create a market for alternating current that eventually became the standard for household use. Ford stuck to producing his successful Model T, giving General Motors and Chrysler the opportunity to establish themselves by supplying a greater variety of cars. The Wright brothers failed as entrepreneurs because they could not abandon their original airplane design to compete with better ones that other engineers soon developed. Edwin Howard Armstrong won a lawsuit against his rival Lee de Forest over a key radio patent. But Armstrong demanded payment that his rival could not make, allowing the case to remain open and so giving later judges the chance to rule in favor of de Forest. The structural designer Othmar Ammann achieved revolutionary economy in the design of the George Washington Bridge, but he also embraced a theory of how the bridge worked that neglected the dynamic effects of wind. The Tacoma Narrows Bridge, designed by another engineer according to the theory, came down in moderate winds in 1940. The difficulties that many engineers faced after their great innovations did not diminish their achievements but reveal the misjudgments that often followed success.

The chapters that follow are primarily about the engineering ideas that helped launch the twentieth century. Each idea was usually simple in its original form; as each new technology began to mature, it became more complex. At the same time, each branch of engineering developed in its own unique way. The networks and machines of the period from 1876 to 1939 propelled their inventors to public acclaim because their work went directly into the homes and garages of ordinary Americans. Process industry was more mysterious; refined in private compounds, shipped to filling stations, and then pumped into cars, gasoline was rarely seen and the engineers who produced it were almost unknown to the outside world. Great bridges were conspicuous landmarks, yet their designers and the innovators in steel and concrete were also largely invisible. Modern engineering often presents two faces: one familiar and iconic, the other anonymous and drab. This book tries to bring the great names down to earth, by making their ideas numerically or conceptually more accessible, while bringing to light the lesser-known engineers whose work still gives shape to modern life today.

CHAPTER TWO

Edison, Westinghouse, and Electric Power

*E*lectric power is indispensable to modern life. Telephones, lights, elevators, and computers depend on electricity, as do the ignition systems of cars and airplanes; without it, society would not have advanced beyond the nineteenth century. Thomas Edison is best known for having invented a new kind of light bulb, but his bulb was not an isolated invention. It was part of a network *network* that Edison engineered to produce and distribute electric power from central generating stations. Following the success of his first power and light network, Edison created generating plants and distribution lines to bring electricity to homes and workplaces that found a wide range of uses in addition to indoor illumination.

Edison used a form of electricity known as direct current, which was practical in places, such as cities, where the users of power were close to the generating plant. Direct current could not be transmitted economically over distances longer than one or two miles. Within a few years, engineers found that a different form of electricity, alternating current, was more practical for longer-distance transmission. Edison could not bring himself to abandon his original technology and embrace alternating current, and a rival entrepreneur, George Westinghouse, seized the opportunity to supply alternating current to users of electric power in competition with Edison. Alternating current has since become the standard for transmission to households and workplaces.

Electricity and Light

Before the nineteenth century Americans relied mainly on candles for light. In 1816 the first U.S. company to provide street lighting with coal gas began operations in Baltimore, processing gas from burning coal and piping it to streetlights. Gas was also piped into urban homes and offices to light indoor lamps. A new petroleum-refining industry arose in the 1850s to supply customers beyond the reach of urban gas networks with kerosene to light indoor lamps (see chapter 4). Indoor gaslight and kerosene were fire hazards, though, and kerosene also produced smoke.[1] The use of electricity for light attracted interest as an alternative.

During the eighteenth and early nineteenth centuries, scientists had learned new principles of electricity and magnetism. In 1800 Alessandro Volta of Italy showed that an electric *current* flowed around a closed wire circuit connected to a battery (sidebar 2.1). An electrical pressure, later named *voltage*, caused the current to flow. Hans Christian Ørsted of Denmark and André-Marie Ampère of France discovered that an electric current flowing through a wire circuit generated a magnetic field around itself. Ampère also found that coiling the wire created a stronger magnetic field inside the coil and made it an electromagnet. In 1825 William Sturgeon of England created a stronger electromagnet by passing current through a wire coiled around a U-shaped iron bar. Georg Ohm in Germany studied the *resistance* of circuits to the flow of current. He discovered a relationship, now known as Ohm's Law, $V = IR$, which stated that voltage (V) is equal to current (I) times resistance (R).[2]

Experiments in the early nineteenth century showed that electricity could be employed to produce light. In 1808 Sir Humphry Davy in England demonstrated an arc light by connecting two carbon rods to a battery. When he touched the rod tips to each other and then separated them, electricity jumped between the tips in a brilliant arc. In the 1840s Sir William Grove ran an electric current through a platinum wire, heating the metal and causing it to incandesce or emit light. Neither arc nor electric incandescent light was practical, though, as long as the only sources of electricity were batteries, which could supply electric power only in small amounts.[3]

In 1831 Michael Faraday of England discovered that rotating an armature in a magnetic field induced a voltage, causing a current to flow in a closed circuit attached to the armature (sidebar 2.2). As long as there was an external source of mechanical power to rotate the armature, the method could generate electric current continuously. This current became known as *alternating current* (A.C.), because with each half revolution of the armature, the voltage reversed direction. Engineers soon invented an attachment (the commutator) that allowed only current in one direction to flow from the armature into the circuit. Current flowing in a single direction became

Sidebar 2.1   **Electricity and Magnetism**

The Electric Circuit

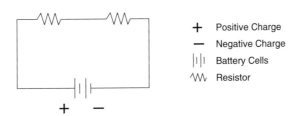

+   Positive Charge
—   Negative Charge
|ı|ı   Battery Cells
⋀⋁⋁   Resistor

Two-cell battery circuit with added resistors (e.g., lamps)

A battery has one or more cells in each of which a negatively charged body (with more electrons than protons) is paired with a positively charged one (more protons than electrons). When a negative and positive body are connected by a conductor (e.g., wire), they form a closed circuit. Electrons flow from the negative body to the positive one.

The pressure on the electrons to flow is called the *electromotive force* or *voltage* (*V*). The electron flow is the *current* (*I*). The size of the current is proportional to the *resistance* (*R*) of the circuit. A lamp offers resistance, as does the conducting wire.

Ohm's Law states that voltage equals current times resistance: $V = IR$. Voltage is measured in *volts*, current in *amperes* or *amps*, and resistance in *ohms*.

Electromagnetism

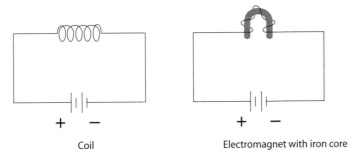

Coil   Electromagnet with iron core

When an electric current flows in a circuit, a magnetic field forms around the current. If the circuit is coiled, the magnetic field will be stronger in the coil by the number of turns per unit of length (e.g., a coil of one foot with three turns will have a magnetic field three times as strong as a line of the same length with one coil).

A straight or horseshoe-shaped iron bar inside a coil will magnetize when current flows through the coil. The bar becomes an electromagnet.

Sidebar 2.2 **Generating Electricity**

Magnetism to Electricity

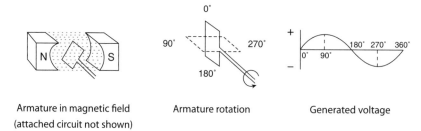

Armature in magnetic field   Armature rotation   Generated voltage
(attached circuit not shown)

A magnetic field forms between the opposite poles of a magnet (*N* and *S*). Rotating an armature (e.g., a wire loop or coil) in the field induces a voltage in the armature and a current in a closed circuit attached to it. A machine that rotates an armature in a magnetic field to produce current is called a *generator*.[1]

Voltage is zero when the armature is at 0° and 180° as shown in the diagram. When the armature is at 90° and 270°, voltage is at a maximum. With each half cycle of the armature (0° and 180°), voltage reverses direction. The first direction is designated as positive (+) and the second direction as negative (−).

Direct and Alternating Current

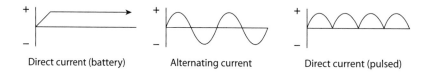

Direct current (battery)   Alternating current   Direct current (pulsed)

The current from a battery is *direct current* (D.C.) because it flows in one direction. Current from a generator reverses direction with each half cycle of the armature and is known as *alternating current* (A.C.).[2] A device called a commutator can reverse the flow of alternating current at each half cycle so that the generator produces only direct current (in pulses). Early generators were called *dynamos*, and alternating current generators were also known as *alternators*.

---

[1] Faraday's original generator, not shown here, had a different rotating element.
[2] In power generation, voltage and current are not normally in phase. Current lags voltage by 90 degrees.

known as *direct current* (D.C.). Feeding back some current to a coil around the magnet, making it an electromagnet, strengthened the magnetic field. Machines that produced electric current became known as generators or dynamos, and engineers soon found uses for the electricity that they could supply.[4]

In 1870 the Belgian Zénobe Théophile Gramme invented a practical dynamo producing direct current with a steam engine to rotate the armature, which the Russian inventor Paul Jablochkoff used in 1878 to power arc lights in Paris, France. The following year Charles F. Brush began to operate systems of arc streetlights in American cities using steam engines and dynamos to produce direct current. Arc lights were far brighter than gaslights and soon became popular for lighting streets. But arc lights were not practical for lighting homes and offices because they were too intense.[5] Thomas Edison believed that the key to indoor illumination was to use incandescent electric light supplied from generators, and he resolved to develop a system of incandescent light and power.

Edison and Incandescent Light

Thomas Alva Edison (1847–1931) learned telegraphy at age fifteen. After working as an itinerant telegraph operator, a growing passion for invention drew him to Boston and then to New York, where he improved the telegraph printers used to transmit gold and stock prices. With money from this work, he launched a business in Newark, New Jersey, making telegraph equipment. In 1876 Edison established a laboratory in Menlo Park, New Jersey, where with a team of assistants he devoted himself to research (figures 2.1 and 2.2). In 1877, Edison invented the phonograph that made him internationally famous. But his greatest achievement was to develop a system to provide incandescent electric light.[6]

Arc lights were connected to each other in series; a single circuit connected the lights and turned them on and off at the same time.[7] However, incandescent indoor lighting would be practical only if each lamp could be turned on and off individually, at different times. To make lights operate in this way, Edison designed a parallel circuit in which, when some lamps were turned off, current would still flow to other lamps that were turned on (sidebar 2.3).

Some influential scientists and engineers argued that such a network could never

Figure 2.1. Thomas Alva Edison in 1881. Source: *"Edisonia": A Brief History of the Early Edison Electric Lighting System* (New York: Association of Edison Illuminating Companies, 1904), p. 78. (Right)

Figure 2.2. Menlo Park laboratory interior, 1880. Courtesy of Edison National Historic Site, West Orange, NJ. Neg. No. 6924.

work. Lamps needed electric *power* (P) to operate, and in 1840 the English scientist James Joule had defined electric power as the voltage times the current, a formula that came to be known as Joule's Law, $P = VI$. Experts in the 1870s assumed that the power available to each lamp on a parallel circuit would diminish by the cube of the number of lamps (see appendix). But Edison realized that a dynamo could supply more lamps by simply generating more power. In the parallel circuit design, voltage had to be constant, so that the voltage supplied was the same to each lamp, whether one light or all of the lights were turned on. Edison designed his system to meet the demand for power at a standard voltage by varying the total current.[8]

The problem facing Edison was the relation of current to resistance expressed in Ohm's Law, $V = IR$. By substituting IR for V, Joule's Law can be rewritten as $P = I^2R$. In a network of incandescent lamps, resistance (R) would consist of two components: the resistance of the filament (the thin material in the lamp bulb that emitted light when heated) and the resistance in the generator and the transmission lines. The power needed to supply a network of lamps would equal the sum of two numbers: the current I^2 multiplied by the filament resistance, and the current I^2 multiplied again by the resistance in the generator and the lines. Although it seemed essential to keep resistance in the lamp filament as low as possible to minimize the product of the resistance and the current squared, such a low resistance would have raised the cost of transmission. The resistance in a transmission line was the product of the resistivity of the conducting material (a fixed value, ρ) and the line length (L), divided by the cross-sectional area (A) of the line (sidebar 2.4). A low resistance in the line would have required a line with a large cross-sectional area, and the best conducting material, copper, was also expensive.

Edison's great insight was to see that if he could instead raise the resistance of the lamp filament, he could reduce the amount of current needed to deliver a given amount of power. The cross-sectional area of the transmission lines could then be much smaller. In his preliminary calculations, Edison estimated that his lighting system could compete with indoor gaslights if it delivered electric power to incandescent bulbs with 100 volts, a current of 1 amp, and a filament resistance of about 100 ohms.[9] (The actual system he built would operate on 110 volts, 0.75 amps, and 147 ohms per bulb.) Arc lights typically operated on a current of 10 amps and their carbon rods had a

Sidebar 2.3   **Series and Parallel Circuits**

Series Circuits for Arc Lamps

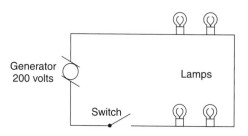

Generator
200 volts

Lamps

Switch

Voltage (V) = 50 volts per lamp
Current (I) = 10 amps
Resistance (R) = 5 ohms per lamps
Number of lamps = 4
Total generated voltage = 200 volts

In a series circuit, lamps are in a single circuit. If one lamp goes off, the circuit breaks and all of the lamps go off. The total current is the same however many lamps are in the circuit, but the total voltage depends on the number of lamps. Lamps can be added to a series circuit as long as the total voltage remains proportional to the number of lamps.

Parallel Circuits for Incandescent Lamps

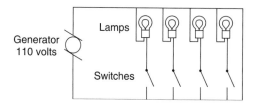

Lamps

Generator
110 volts

Switches

Voltage (V) = 110 volts
Current (I) = 0.75 amps per lamp
Resistance (R) = 147 ohms per lamp
Number of lamps = 4
Total generated current = 3 amps

In a parallel circuit, lamps connect so that if one lamp turns off, current is available to other lamps that continue to operate. Any number of lamps can be on or off. With lamps that each operate at a standard voltage, the voltage at each lamp must be the same, and total voltage is therefore constant. But the total current depends on the number of lamps. Lamps can be added to a parallel circuit as long as the total current is proportional to the number of lamps.

Source: Harold C. Passer, *The Electrical Manufacturers, 1875–1900* (Cambridge, MA: Harvard University Press, 1953), p. 81. Voltage in the Edison system varied within a margin of 2 percent, an acceptable amount. Transmission losses are neglected in the diagrams above.

Sidebar 2.4   **High Resistance and Low Cost**

Ohm's Law and Resistance

Ohm's Law states that $V = IR$ (voltage = current × resistance)

Resistance R can be written as $\rho l/A$, where

ρ = resistivity of conductor (0.00000067 ohm-inches for copper)
L = length of line (in inches)
A = cross-sectional area of line (in square inches)

Upton's Calculations

Thomas Edison's assistant, Francis Upton, made calculations to demonstrate Edison's insight that incandescent electric lamps needed to have high resistance. Using Ohm's Law and the formula for line resistance, Upton calculated the cross-sectional area (and thus the amount of copper) required in a transmission line of 12,000 inches to supply a current of 1 amp at 100 volts to lamps with a high resistance of 100 ohms:

$$I = \frac{V}{R} = \frac{V}{\rho l / A}$$

$$A = \frac{I(\rho L)}{V} = \frac{(1)\,\rho\,(12{,}000)}{100}$$

$$A = 120\,\rho \text{ square inches}$$

Upton then calculated that if the lamps operated with a low resistance of 1 ohm and a current of 10 amps at 10 volts over the same line length, the cross-sectional area of the line would have to be 100 times greater:

$$A = \frac{I(\rho L)}{V} = \frac{(10)\,\rho\,(12{,}000)}{10}$$

$$A = 12{,}000\,\rho \text{ square inches}$$

The advantage of high resistance was clear. Edison would design each of his lamps to operate on a current of 0.75 amps at 110 volts with a resistance of 147 ohms.

Source: Francis Jehl, *Menlo Park Reminiscences* (Dearborn, MI: Edison Institute, 1936–41), 2:852–54, 3:1077–79.

resistance of about 5 ohms.[10] To be economical, Edison's system required a low current, which in turn depended on finding a lamp filament with high resistance.

Filaments tested by earlier inventors burned out quickly and Edison at first tried to find a longer-lasting metal. In September 1878 he announced that he would soon have a system of incandescent light on the assumption that platinum would be a durable filament. The following year, though, he found that heated platinum released trace gases that caused the filament to burn out too quickly for practical use. Edison eventually decided that a carbonized thread would work as a high resistance filament in a bulb with a high vacuum. A device to create such a vacuum, the mercury pump, had been invented in the mid-1860s, and Edison improved it for his own use. Over the night of October 21–22, 1879, he and his staff tested a vacuum bulb containing a carbon filament with a resistance of more than 100 ohms. The bulb gave the desired illumination (16 candlepower) for $14\frac{1}{2}$ hours on 1 amp of current (figure 2.3). Edison filed a patent for the bulb on November 1, 1879.[11]

In developing his bulb, Edison differed from the approach taken by Sir Joseph Swan (1828–1914), an English chemist who had taken an interest in incandescent light as early as 1848. Swan resumed his light research in 1877, and in early 1879 he demonstrated an incandescent bulb. But Swan designed his bulb to have a low resistance, and he did not develop a practical filament until 1880. His approach was to demonstrate the bulb as an isolated laboratory experiment; the needs of a commercial system played no role in his thinking. To avoid protracted litigation, though, Edison reached an agreement with Swan in 1882 to share the British market.[12]

The parallel circuit and the high-resistance filament were crucial to Edison's system, but he needed to solve two further problems before he was ready to build it. One was related to distance. A constant voltage would eventually diminish the farther it went from the generator. To offset the line loss associated with increasing distance, Edison would have needed to increase the cross-sectional area of his transmission lines, increasing their cost. Instead, he designed shorter feeder lines to go out from the power plant. Over these lines voltage would fall from 120 to 110 volts. The feeders would connect to service mains and lamp circuits that could operate on 110 volts. Besides distributing power more efficiently, the system of feeders and mains reduced to one-fifth the amount of copper that Edison believed his lines would otherwise require.[13]

Figure 2.3. Replica of
1879 Edison light
bulb. Courtesy of
General Electric
Archives, now the GE
Photograph Collection,
Schenectady Museum,
Schenectady, NY.

Finally, Edison needed a more efficient dynamo to generate electric power. Scientists and engineers in the 1870s assumed that the internal resistance in a dynamo had to be in balance with the external resistance of the circuit and the lamps on it, because a battery achieved the most efficient output of power when internal and external resistances were equal. But with a dynamo, balancing the resistances meant losing half of the energy in the generator itself. In contrast to his lamps, Edison realized that he would need to have a dynamo with very low internal resistance (sidebar 2.5). No existing

Sidebar 2.5 **The Edison Dynamo**

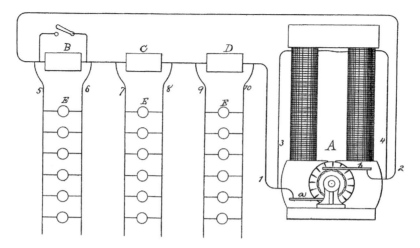

At Edison's Pearl Street station in New York City, each dynamo supplied about 1,200 lamps with power equal to 110 volts and 0.75 amps at each lamp:

$$P = VI = (110)(0.75)(1200) = 99,000 \text{ watts or } 99 \text{ kW (kilowatts)}$$

A feeder line with a cross-sectional area of one square inch connected each dynamo to service mains that fed smaller wires to the lamps. To cover a loss of $R = 0.008$ ohms per 1,000 feet in the feeder line (neglecting losses in the mains and lamp circuits), the dynamo produced an additional 6.48 kW:

$$P = I^2R = (900)^2 (0.008) = 6.48 \text{ kW} + 99 \text{ kW} = 105.5 \text{ kW}$$

To offset voltage loss in the feeder line, each dynamo generated power at 120 volts (in the preceding calculations, 105.5 kW/900 amps = 117 volts), which fell to 110 volts at the lamps.

Edison designed his dynamo with a very low internal resistance, $R = 0.0039$ ohms, requiring an additional 3.54 kW in the dynamo, for a total of 109 kW. The dynamo therefore had an efficiency of 105.5/109 kW, or 96 percent.

If Edison had designed his dynamo with internal and external resistances equal, an efficiency of only 50 percent, he would have lost 105.5 kW in the dynamo itself as heat in order generate the 105.5 kW needed by the lines and lamps.

Sources: Diagram from T. A. Edison, U.S. Patent No. 379,772 (1888). Data from Francis Jehl, *Menlo Park Reminiscences*, 3 vols. (Dearborn, MI: Edison Institute, 1936–41), 3:1077–78. For voltage drop in the feeders, see 2:822. For internal resistance in the Edison dynamo, see Charles L. Clarke, "Edison 'Jumbo' Steam Dynamo," in *'Edisonia': A Brief History of the Early Edison Electric Lighting System* (New York: Association of Edison Illuminating Companies, 1904), p. 41.

generator could meet this need, so while they worked on the other elements of their system, Edison and his team designed a new dynamo with a low internal resistance that generated direct current with an efficiency of more than 90 percent (figure 2.4).[14]

As the location of his first network, Edison chose the center of American finance, the Wall Street district of New York City. He calculated that 10,000 gaslights in the district cost $136,875 to operate four hours per day. Edison believed that he could supply a similar number of incandescent electric lamps giving the same luminosity for $90,886 per year, with a start-up cost of $150,680. The cost of launching his system proved to be twice this amount, but the success of his bulb in 1879 enabled Edison to raise the money from investors led by the banker J. P. Morgan. On September 4, 1882, the network began operation from a central generating plant on Pearl Street, where six steam engines turned six "Jumbo" dynamos, so named by Edison for the elephant featured in the P. T. Barnum traveling circus. Each dynamo supplied power to about 1,200 lamps, and each lamp received 110 volts and 0.75 amps and had a filament resistance of about 147 ohms. Electric power is now measured in watts, and each lamp used 82.5 watts (110 volts times 0.75 amps). To measure usage, Edison developed an electrochemical meter and then charged by the kilowatt-hour (1,000 watts per hour).[15]

Edison soon launched companies in other states and countries to make power equipment, manufacture light bulbs, and provide electric utility service. His American firms merged in 1889–90 to form the Edison General Electric Company.[16] In addition to supplying lamps, direct current from generators made electric motors practical. These worked as generators in reverse: direct current supplied to an armature inside a magnetic field caused the armature to rotate, producing mechanical power. Electricity provided industry with a new way to drive machines.[17]

Westinghouse and Alternating Current

Thomas Edison's system worked well in urban and industrial areas where power plants could be located close to consumers. But Edison's network could not extend over longer distances because voltage (and thus power) losses increased with the length of transmission lines. To prevent the dimming of lights with distance, Edison devised

Figure 2.4. Pearl Street station dynamo room. Source: *Scientific American*, 47:9 (August 26, 1882): 127.

ways to branch and wire his system more efficiently. But he found that he could not transmit power economically over a distance of more than about two miles.[18]

George Westinghouse (1846–1914) found a way to transmit electric power over longer distances by switching from direct to alternating current. Born near Schenectady, New York, Westinghouse worked in his father's machine shop and invented a device for putting derailed train cars back on track. After moving to Pittsburgh in 1868, he studied the problem of braking locomotives, a growing hazard as hand-turned brakes became more difficult to operate on heavier trains. In 1869 Westinghouse invented the

Figure 2.5. George Westinghouse. Source: Francis E. Leupp, *George Westinghouse: His Life and Achievements* (Boston: Little, Brown, 1918), frontispiece.

compressed air brake and formed a new company to sell air brakes to the Pennsylvania Railroad and other rail lines. This invention made him wealthy (figure 2.5).[19]

Westinghouse also discovered natural gas under his Pittsburgh estate that was under high pressure. The gas had commercial value but only if the pressure could be reduced for delivery to consumers. Westinghouse devised a system that sent the high-pressure gas long distances through relatively inexpensive small-diameter pipes. At customer locations, he switched to large-diameter pipes, transforming the flow from high pressure to low pressure. This idea of high-pressure transmission and low-pressure usage helped him see the value in a new mode of electrical power supply, when he read an article in 1885 about an English system that used alternating current to supply power to incandescent lamps.[20] The key to the new system was the use of *transformers*.

A transformer (sidebar 2.6) can be thought of as a ring-shaped core of magnetizable material, usually iron or steel, with two separate wire circuits, one coiled around each side of the ring core. Engineers call the first wire the primary circuit and the other the secondary, and normally each circuit is closed. With power equal to voltage times current, $P = VI$, we can designate the power in the primary circuit V_1I_1 and the power in the secondary circuit V_2I_2. When the power V_1I_1 begins to flow, it induces a magnetic current or flux in the ring core. The change in flux induces a voltage V_2 that causes a current I_2 to flow in the secondary circuit.

Direct current is not practical for use in a transformer because a current going in one direction can induce only a momentary magnetic flux in the ring core. However, an alternating current induces a flux in the ring core each time it changes direction, and the continuous changes enable an alternating current to sustain itself in a closed secondary circuit. A transformer can also reduce or increase the voltage between the two circuits according to the ratio of turns that the two wires make around the ring. As shown in sidebar 2.6, the ratio of primary to secondary turns is one to twenty. If we assume no power losses in the transformer, 50 volts and 800 amps in the primary circuit will transform into 1,000 volts and 40 amps in the secondary. A transformer can in this way step up the voltage and step down the current for long-distance transmission, so that line losses (proportional to I^2 in the line loss $P = I^2R$) can be kept low. A transformer at the other end can then step the voltage back down and the current back up for domestic and industrial use.[21]

Westinghouse realized that he could transmit electric power over much longer distances than Edison by using alternating current and transformers. After studying some transformers that he ordered from England in late 1885, he paid William Stanley, an electrical engineer in Great Barrington, Massachusetts, to develop a better transformer. In January 1886 he formed the Westinghouse Electric Company and tested Stanley's transformers the following autumn with incandescent bulbs that Westinghouse designed. An alternating current generator produced 800 amps at 50 volts that

Sidebar 2.6 **Transformers and Alternating Current**

Generator	Transformer A	High Voltage Lines	Transformer B	Lights
$P_1 = V_1 I_1$		$P_2 = V_2 I_2$		$P_1 = V_1 I_1$
$V_1 = 50$	$N_1 = 1$	$V_2 = 1{,}000$	$N_2 = 20$	$V_1 = 50$
$I_1 = 800$	$N_2 = 20$	$I_2 = 40$	$N_1 = 1$	$I_1 = 800$

A transformer can step up a voltage and step down a current for the long-distance transmission of electric power. A transformer at the other end can step the voltage back down and the current back up for local use.

If power losses in the transformers A and B are negligible, power in the primary circuit (P_1) will equal power in the secondary circuit (P_2). With $P = VI$, $V_1 I_1 = V_2 I_2$ and $I_2 = I_1 \, (V_1/V_2)$. Voltage between the primary and secondary circuits transforms according to the ratio of primary to secondary turns (N_1/N_2) that the two circuit lines make around the transformer core.

In his 1886 Pittsburgh to Laurenceville test, Westinghouse transformed 800 amps at 50 volts to 40 amps and 1,000 volts for transmission. The ratio of N_1/N_2 was therefore one to twenty. Transformers at the end of the line stepped the voltage back down and distributed the 800 amps to 400 lamps, giving each lamp 50 volts and 2 amps of current.

Source: Harold C. Passer, *The Electrical Manufacturers: 1875–1900* (Cambridge, MA: Harvard University Press, 1953), p. 138. In the Laurenceville test, some additional power (neglected here) offset line loss.

stepped up to 1,000 volts and 40 amps for transmission from Pittsburgh to the town of Laurenceville, four miles away. Transformers there stepped the voltage back down and distributed it to 400 lamps, giving each lamp 50 volts and 2 amps.[22]

Edison's bulbs did not operate with the same volts and amps, but the difference in efficiency can be seen by comparison: to power 400 of Edison's bulbs, at 110 volts, an

Edison line would have had to carry a current of 300 amps. Unable to use transformers, the Edison line would have needed more than fifty-six times as much copper wire as the Westinghouse line to transmit this current over the same four miles with the same line loss. If Westinghouse had stepped up his transmission voltage to 10,000 volts instead of 1,000, his alternating current could have traveled 400 miles with the same line loss.

Westinghouse soon received orders for A.C. generating systems. By the end of 1890, he had sold 350 central station alternating-current systems to challenge the Edison General Electric Company's 400 direct-current central stations.[23] Edison bitterly fought alternating current in the 1880s, in what became known as the "Battle of the Currents," in part because he was unwilling to abandon the system that had earned him success and in part because he believed that alternating current was dangerous. Edison tried to persuade state legislatures to limit transmission to several hundred volts, which would have deprived alternating current of its economy, and to discredit A.C. further by emphasizing its risks, he supported its use to execute criminals.[24] But Westinghouse showed that the alternating current could be safe for local use, and he seized the opportunity that Edison gave him to create a new market, especially in areas where direct current was not economical. Light bulbs could work on alternating current, because at sixty cycles per second the flicker was too rapid to be noticed by the human eye.

The Edison company needed to offer A.C. equipment to remain competitive, and in 1892, Edison General Electric merged with a rival firm, Thomson-Houston, which had alternating-current technology. The merger created the General Electric Company (GE).[25] Edison left the business and moved on to other interests, such as motion pictures, while Westinghouse continued to manage his own company. Westinghouse scored a public relations triumph by illuminating the 1893 Columbian Exposition in Chicago. GE and Westinghouse reached a patent-sharing agreement in 1896 and thereafter dominated the manufacturing of electrical equipment in the United States.[26]

Incandescent light did not cause gas and arc light to die out at once, and kerosene lamps continued to be needed in remote areas until rural electrification arrived. But gas, arc, and oil light eventually declined in the twentieth century. The gas companies found a new market supplying piped gas to kitchen cooking ranges and home heaters.[27] Electric utilities eventually became separate publicly regulated companies, leaving

GE and Westinghouse to manufacture electrical goods. As the twentieth century advanced, electricity became an increasingly vital part of modern life.[28]

Engineering Research and Practice

As the electrical industry matured, it began to rely on engineers with more formal training in mathematics and science. With alternating current, a number of complex technical challenges had to be overcome for the technology to be used more efficiently and more widely. A key problem was the lack of an electric motor able to work on alternating current. In a D.C. motor, supplying direct current to an armature inside a static magnetic field caused the armature to rotate. Alternating current, however, produced no rotation. The armature tried to rotate in alternate directions, causing each motion in one direction to be canceled by the motion in the other.

Nikola Tesla (1856–1943) invented a motor that solved this problem by rotating the magnetic field. Tesla had studied engineering in Austria and had emigrated from his native Serbia to the United States in 1884 (figure 2.6a). Three years later, he patented an electric motor in which an armature stood at the center of four coils arranged in a ring at ninety-degree angles to each other. An alternating current went to one opposite pair of coils. When the current reached its maximum, a second alternating current went to the other opposite pair. The currents pulled the armature magnetically and caused it to rotate. In addition to working on alternating current, the motor did not require metal contacts between the circuits and the armature and thus avoided the sparking that occurred with such contacts in direct-current motors. Tesla patented other inventions but sold his patents early in his career. He failed as an entrepreneur and died in tragic obscurity.[29]

A second problem was magnetic loss in transformers. Charles Steinmetz (1865–1923) solved this difficulty and went on to set new standards in the design of alternating-current technology as an engineer for General Electric (figure 2.6b). Born in Germany, he nearly completed a doctorate in mathematics when he had to flee the country in 1888 because of his socialist views. After engineering study in Zurich, Switzerland, he arrived in New York City and found work in a small firm that made electrical equipment. He soon turned his attention to the problem of magnetic loss. Just as electric current had to overcome resistance in a circuit, so did the magnetic flux have to overcome

Figure 2.6a. Nikola Tesla. Courtesy of
the Prints and Photographs Division,
Library of Congress, Washington, DC.
LC-USZ62-61761.

Figure 2.6b. Charles Steinmetz. Courtesy of
General Electric Archives, now the GE
Photograph Collection, Schenectady Museum,
Schenectady, NY.

a resistance (called reluctance) in a metal transformer. In 1890 Steinmetz devised a formula to measure this loss so that engineers could design transformers to work more efficiently.[30] General Electric acquired his employer in 1893, and the following year Steinmetz moved to the main GE plant in Schenectady, New York, to head its calculating department. Here he developed an approach to the design of alternating-current circuits that provided engineers with a practical way to understand them.[31]

Steinmetz made his contributions at a time when engineering was beginning to develop a more professional identity. In the late nineteenth century, universities began to establish new engineering departments, and engineers began to organize professional societies. In academic engineering, research and theory came to be emphasized and rewarded more than practical design experience. Modern physics used calculus, and in the 1890s some engineers tried to describe and solve engineering problems in the manner of physics problems. Their goal was to give engineering a more theoretical basis that would make design largely a matter of deduction from theory. Steinmetz shared the spirit of modern science in its search for general principles and exact methods. But he and other engineers in the electrical industry also recognized that the objects of engineering design had unique characteristics and that these also called for empirical understanding and guidance. Steinmetz and his colleagues in industry had to balance the academic ideal of open intellectual exchange with the proprietary need of industry to keep much of its knowledge secret. But Steinmetz was able to share his most important ideas, which were not proprietary, with the profession.[32]

Steinmetz's success in developing standard formulas and methods of design ironically enabled General Electric engineers to do without him, and after 1900 he moved into the role of a consultant. At the start of the twentieth century, two new types of incandescent lighting, a metallic filament lamp invented in Germany and a mercury vapor lamp pioneered in New York City, began to compete with Edison light bulbs. In response, Steinmetz persuaded General Electric to establish a research laboratory in September 1900, where William David Coolidge (1873–1975) developed a method for making tungsten wire malleable enough to use in filaments. Tungsten is still used in light bulbs today. Irving Langmuir (1881–1957), a chemist who joined the General Electric Laboratory in 1909, extended the life of the tungsten filament bulb and for other work won the 1932 Nobel Prize in chemistry.

These researchers approached problems by seeking first to understand the basic physical and chemical facts and then by looking for ways to meet engineering needs. Their work exemplified the greater use of scientific knowledge to solve the more incremental problems that industry tended to have once its basic technology and business were established. The success of the General Electric Laboratory inspired the Westinghouse Company to establish a laboratory with chemists and physicists in 1916. General Motors incorporated Charles Kettering and his laboratory into its organization in the early 1920s, and the American Telephone and Telegraph Company created a new research institution, the Bell Telephone Laboratories, in 1925.[33]

The electrical discoveries of the early nineteenth century made it possible for Thomas Edison to create his network of light and power. But the science of his time did not encourage such innovation. By the 1870s, leading scientists and theoretically-minded engineers had concluded that the sub-division of electricity for light was impossible, and Edison would have never achieved his great innovation if he had been guided by such judgment. He was no tinkerer nor mere trial-and-error experimentalist; his novel conception of high resistance in the filament and low resistance in the dynamo proceeded from his radically new insight into the use of Ohm's and Joule's Laws. Edison achieved his breakthrough because he thought as an electrical engineer, not as an applied scientist.

After alternating current came into more general use, engineers with more formal academic training in mathematics and science then worked out a more theoretical grounding of the phenomena involved and made the new technology more efficient. Engineers such as Steinmetz recognized, however, that their work differed from science and that electrical engineering was not simply a matter of applying theory to practice.

CHAPTER THREE

Bell and the Telephone

As Edison was beginning his work at Menlo Park, the telephone made its first public appearance at the Philadelphia Centennial Fair on June 25, 1876. Its inventor, Alexander Graham Bell, envisioned a network of telephones that would enable any person to reach any other person in the world instantly. Bell's ambition has yet to be fully realized but in the twentieth century the telephone became indispensable to daily life in the United States. Before the phone, voice communication was limited to distances measured in feet. Telegraph messages traveled longer distances but had to be encoded and decoded by operators at each end. With a telephone, communication was direct, and no place on the network was more than a simple phone call from any other place on it.

network

Bell did not set out to invent a telephone; his original goal was to invent a better telegraph. But he soon realized that his approach to telegraphy might make voice communication possible. After patenting his telephone and then demonstrating it, he formed a company to provide telephone service and then retired from the business. As the network grew, it encountered technical difficulties, principally the attenuation or weakening of calls over longer distances. To overcome this challenge, the Bell System turned to more scientifically trained engineers, who achieved long-distance telephony and met the growing need for carrying capacity as the number of telephone subscribers grew.

Bell to Boston

Born in Edinburgh, Scotland, Alexander Graham Bell (1847–1922) grew up with the ambition to follow the profession of his father, Alexander Melville Bell (1819–1905), a well-known speech teacher at the University of Edinburgh.[1] Melville Bell's invention of "visible speech" taught deaf people how to communicate vocally (figure 3.1a and b).[2]

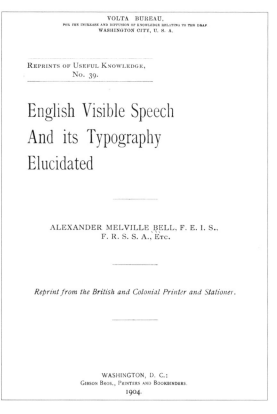

Figure 3.1a. Alexander Melville Bell.
Source: Alexander Melville Bell,
English Visible Speech and Its Typography Elucidated (Washington, DC: Volta Bureau, 1904), frontispiece.

Figure 3.1b. Title page of Bell's book on visible speech.
Source: Alexander Melville Bell, *English Visible Speech and Its Typography Elucidated* (Washington, DC: Volta Bureau, 1904).

Using this approach, his three sons worked with him as teachers of the deaf. The middle son, Alexander Graham, also took an interest in the mechanics of sound. He found that he could produce vowel sounds with tuning forks, and he learned of an experiment in which the German scientist Hermann von Helmholtz had used the magnetic field of an electric current to resonate tuning forks. Bell mistakenly believed Helmholtz had demonstrated that sound could be carried by electricity. But the misunderstanding encouraged Bell to pursue the idea that speech might be carried by an electric current.[3]

Melville Bell's first and third sons died of tuberculosis as young men, and the failing health of Alexander Graham prompted the family to move to Canada in 1870. The Bells settled in Brantford, Ontario, seventy-five miles west of Niagara Falls. The younger Bell soon recovered and accepted an invitation to teach at a school for the deaf in Boston, Massachusetts, where he arrived in April 1871. His success teaching deaf students to communicate soon attracted notice. In the fall of 1873, he joined the faculty of the newly founded Boston University as a professor of vocal physiology. To save money, he moved to Salem, just outside the city, where he lived with the family of Thomas Sanders, a merchant. Bell taught his deaf son, George Sanders, in return for room and board and commuted to his teaching in Boston.

America's first integrated textile factories had begun outside Boston in the 1820s, and by the 1870s the city was a leading center of telegraph research. The wealth of its industry had enhanced the city's role as a great center of learning, with Harvard College and the newer institutions of Boston University and the Massachusetts Institute of Technology in the forefront. An invitation to lecture at MIT on visible speech and a tour of the school's laboratories renewed Bell's earlier interest in the relationship between electricity and sound.[4] In the spring of 1875 he reduced his teaching (and his income) in order to devote more time to private research (figure 3.2).

From Telegraph to Telephone

The discovery of electromagnetism in the early nineteenth century prompted some researchers to investigate its possible use in telecommunication. In 1830 the physicist Joseph Henry demonstrated how electricity and magnetism could be employed to transmit a signal. Henry wrapped an insulated copper wire around an iron U-shaped electromagnet. Inside the arms of the magnet stood one end of a permanently magnetized steel bar on a pivot. When Henry closed the wire circuit, sending battery current around the magnet, the magnetic force pulled the inside end of the steel bar. The outside end of the bar swiveled and struck a bell (sidebar 3.1).[5]

Henry saw his bell experiment as a way to demonstrate a scientific principle; he had no interest in its practical use for communication. The development of a practical telegraph was the work of Samuel F. B. Morse (1791–1872), who devised an electric circuit

Figure 3.2. Alexander Graham
Bell, circa 1875. Courtesy of
AT&T Archives, Warren, NJ. No.
H-1/8. Property of AT&T Archives.
Reprinted by permission of AT&T.

powered by batteries (sidebar 3.1). Pressing a sender key interrupted the flow of current, causing a printer on the receiving end to print in a code of dots and dashes that
Morse devised to transmit information. Morse patented his telegraph in 1840 and
demonstrated it in 1844 over a line erected by Ezra Cornell from Baltimore to Washington. Telegraph companies soon began wiring the United States under license from
Morse. Cornell helped found Western Union, which became the dominant U.S. telegraph company after the Civil War. As telegraphy spread, operators found that they
could transmit messages faster if the signal activated a sounder instead of a printer.[6]

Sidebar 3.1   **Electromagnetism and Telegraphy**

Henry and Electromagnetism

In 1830 Joseph Henry demonstrated the principle of an electromagnetic telegraph. When he closed a wire circuit coiled around a horseshoe electromagnet (by connecting the two loose wire ends to a battery), the magnetism attracted one end of a magnetized steel bar. The other end swung and sounded a bell.

The Morse Telegraph

In 1844 Samuel F. B. Morse used electromagnetism to telegraph a message from Baltimore to Washington. In the simplified circuit depicted above, pressing key *K* demagnetized the two electromagnets

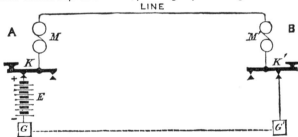

Simplified Morse telegraph circuit

M and *M'*, activating a printer or a sounder (not shown) at *K'*. Grounding each end of the line in the earth was sufficient to complete the circuit.

Sources: Joseph Henry, *The Scientific Writings of Joseph Henry* (Washington, DC: Smithsonian Institution, 1886), 2:434; Frank L. Pope, *Modern Practice of the Electric Telegraph* (New York: D. Van Nostrand, 1872), pp. 25 (circuit).

During the mid-nineteenth century, as railways and steamships brought the world closer together, electric telegraphy over land lines and undersea cables provided nearly instantaneous transmission of news and messages. By the 1870s, however, the volume of telegraph traffic had grown so much that congestion had become a serious problem. Engineers began looking for ways to multiplex messages, to send more than one message over the same line at the same time. In 1872 Joseph Stearns of Boston patented a "duplex" telegraph with the ability to send a message in each direction simultaneously, while Thomas Edison developed a "quadruplex" system in 1874 that could transmit two messages in each direction over a single line at once.[7] Fortune appeared to await inventors who could send even more messages. Alexander Graham Bell began his career as an inventor by joining the search for a better multiplex telegraph.

Bell's approach reflected his interest in hearing and sound. He knew that the human eardrum vibrated to sounds carried through the air. He also knew that the magnetic field of an electric current could vibrate a tuning fork and produce sound. He wondered if sound could be carried by an electric current. When current activated an electromagnet, a magnetic field or *flux* formed around the magnet. Bell's idea was to place a tuned steel reed close to an electromagnet and tap it. The sound vibration in the air would vary the flux and, in so doing, vary the electric current as well. He believed that the current would transmit the variation to the flux of a distant electromagnet on the same circuit, and an adjacent steel reed tuned to the same frequency would then sound (sidebar 3.2). With a set of tuned reeds at one end, Bell hoped that he could send multiple messages over a single line at the same time to a set of duplicate reeds at the other end. The vibrations would combine for transmission. If the receiving reeds could separate them, by each vibrating to one tone, a large number of messages could be sent and received simultaneously.[8]

To develop a "harmonic" telegraph, Bell needed financial backing. In 1873, at the invitation of Gardiner Hubbard, a prominent Boston attorney, Bell had begun to teach Hubbard's deaf sixteen-year-old daughter, Mabel (figure 3.3a, b). Hubbard, who had a keen interest in telegraphy, opposed the monopoly of Western Union on long-distance communication. Bell's idea promised a way to send more messages less

Sidebar 3.2   **Bell's Harmonic Telegraph**

From Mechanical Action to Mechanical Resonance

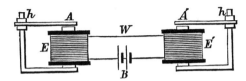

In conventional telegraphy, electromagnetism activated a mechanical striking action in the receiving apparatus. In Bell's harmonic telegraph (above), electricity and magnetism instead produced a mechanical resonance or sound.

An operator sounded a tuned steel reed *A* over electromagnet *E*. The vibrations of *A* varied the magnetic flux at the gap between *A* and *E*, fluctuating the electric current between electromagnets *E* and *E'* and transmitting the fluctuation to the magnetic flux around *E'*. The receiving reed *A'*, tuned to *A*, was to vibrate in resonance.

Multiplexing: Many Messages at Once

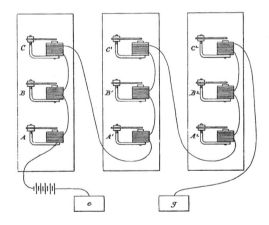

In duplex and quadruplex telegraphy, two or four messages could travel at the same time. In Bell's harmonic telegraph, a set of tuned reeds (*A*, *B*, and *C*) would send a group of tones along a single wire. If the receiving reeds separated them, many more messages could be sent over a single line at the same time and to more than one station (reeds A^1 and A^2 would resonate with reed *A*, etc.).

Sources: George B. Prescott, *Bell's Electric Speaking Telephone* (New York: D. Appleton, 1884), p 70; Alexander Graham Bell, U.S. Patent No. 174,465 (1876).

Figure 3.3a. Gardiner Hubbard. Courtesy of AT&T Archives, Warren, NJ. No. 95-1029. Property of AT&T Archives. Reprinted by permission of AT&T.

Figure 3.3b. Mabel Hubbard. Courtesy of AT&T Archives, Warren, NJ. No. 88-200499. Property of AT&T Archives. Reprinted by permission of AT&T.

expensively.[9] Together with Thomas Sanders, Hubbard formed a partnership with Bell in February 1875 to support the latter in the development of a harmonic telegraph. Bell agreed to give all three joint ownership of the patent rights. To help him in his research, he found a capable technician, Thomas A. Watson, from the workshop of Charles Williams in Boston, where Thomas Edison had also had equipment made.[10]

Bell found that he could not tune his harmonic telegraph precisely enough to eliminate mutual interference of the reeds with each other at the receiving end. His difficulty prompted him to think more deeply about electricity and sound. The advantage of his idea was its use of varying frequencies and amplitudes to transmit information without interrupting the current. He believed that more signals in Morse code could travel in this way at a given time than as interruptions in the current. Bell soon realized that a continuous current was poorly suited to sending messages in a code of stops and starts. A continuous current that conveyed varying frequencies and amplitudes was ideally suited, however, to carrying the human voice (sidebar 3.3).

Earlier inventors had tried to use an electric current for telephony without fully understanding the need for the current to be continuous. In 1861 Philip Reis in Germany placed a metal surface or diaphragm at each end of a telegraph line. Speaking onto the diaphragm vibrated it, and the receiving metal surface reproduced the frequency (high or low tone) of the sound at the other end. But the Reis telephone circuit only closed when the diaphragm vibrated. It acted like a telegraph, in stops and starts, and could not transmit changes in the amplitude (intensity or loudness) of the sound.[11] Bell understood that a telephone needed to transmit varying amplitudes as well as frequencies to carry intelligible speech and that to do so required an uninterrupted current. Bell received encouragement from Joseph Henry, who had served as the secretary of the Smithsonian Institution since 1846. The almost eighty-year-old Henry met the twenty-eight-year-old Bell in March 1875 and listened to his idea of using an "undulatory" electric current to carry voice. "You have the germ of a great invention," Henry said. When Bell confessed his need for more knowledge about electromagnetism, Henry told him firmly, "Get it."[12] Bell redoubled his study.

A breakthrough finally came on June 2, 1875. While testing the reeds of a harmonic telegraph, Bell and Watson found that one of the reeds didn't sound, so they turned off the battery current. Thinking it was stuck, Watson plucked the reed. Enough residual magnetism was in the circuit to cause a receiving reed to sound in another room where Bell heard it. The sound was not the normal tone of the receiving reed. Bell realized with excitement that the circuit had transmitted a new sound. The next day, he spoke onto a diaphragm connected to a reed and failed to produce speech in an identical receiver in another room. But he was now convinced that he was on the right track.[13]

Sidebar 3.3 **Signals and Sound**

Frequency and Amplitude

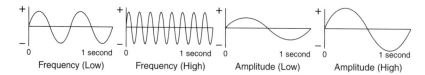

Signals travel electrically as currents that vary in frequency and amplitude. Frequency is the number of cycles per unit of time (usually one second). High and low amplitudes signify the intensity of the signal.

From Intermittent Taps to Continuously Varying Tones

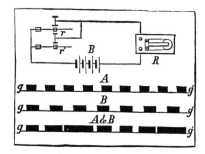

"Make and break" signal

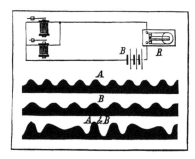

Bell's "undulatory" current

Conventional telegraphy sent messages as "makes" and "breaks" in the circuit. Crowding out eventually limited the number of signals possible to send over a telegraph at any one time.

Bell's idea was to transmit signals in a continuous current that varied in frequency and amplitude. The sum of tones formed a complex waveform of alternating current. Additional tones did not crowd out those already there. Adding an alternating current to a direct current created what Bell called an "undulatory" current (now called a *fluctuating current*). Bell hoped to separate the frequencies in his receivers. Eventually, he realized that he could leave the complex waveform intact and use it to transmit voice.

Source: George B. Prescott, *Bell's Electric Speaking Telephone* (New York: D. Appleton, 1884), pp. 62 (make and break), 64 (undulatory current).

Hubbard and Sanders disapproved of Bell's interest in telephony, which they regarded as a distraction. There was a proven market, in their view, only for a new kind of telegraph. To complicate matters, Bell began to realize that his feelings toward his deaf student, Mabel Hubbard, were changing from teacher to suitor. At age twenty-eight he was eleven years older, and her parents objected. But on August 26 Bell had a meeting with Mabel and believed that all was not lost. Just before Thanksgiving, Gardner Hubbard demanded that Bell concentrate on telegraphy or lose contact with Mabel. Mabel settled the issue on Thanksgiving Day, her eighteenth birthday, by declaring her love for Bell and asking her father for the engagement. Hubbard eventually accepted both the potential son-in-law and his telephone.[14]

On January 20, 1876, Bell notarized a patent application that Hubbard's attorneys in Washington, DC, filed on February 14. The Patent Office granted the claim on March 7, 1876, as Patent No. 174,465. This patent, the first of his two telephone patents, consisted of two pages of drawings and four pages of accompanying text. Bell argued that a conventional telegraph line had a limit to the number of messages it could carry at any one time with an interrupted current. He then explained how a harmonic telegraph and a continuous undulatory current would overcome this problem. Then he noted: "I desire here to remark that there are many other uses to which these instruments may be put, such as the simultaneous transmission of musical notes, differing in loudness as well as in pitch, and the telegraphic transmission of noises or sounds of any kind." Bell then outlined his revolutionary idea, a telephone, almost as an afterthought, and claimed for it the ability to transmit the human voice.[15]

Bell's demonstration at the Philadelphia Centennial in June 1876 drew national attention to his invention, and over the following year he lectured widely on it (figure 3.4). On July 9, 1877, Hubbard, Bell, Sanders, and Watson formed the Bell Telephone Company, and two days later Bell and Mabel Hubbard were married.

Figure 3.4. Bell's 1877 lecture in Salem, Massachusetts. Source: *Scientific American*, 36:13 (March 31, 1877): 191.

Figure 3.5. Elisha Gray. Courtesy of AT&T Archives, Warren, NJ. No.
W11-1. Property of AT&T Archives. Reprinted by permission of AT&T.

Bell, Gray, and Edison

Soon after filing his patent application, Bell learned that a rival inventor had filed a provisional claim (called a *caveat*) for a telephone later the same day, February 14, 1876. The rival was Elisha Gray (1835–1901), a telegraph engineer (figure 3.5). In 1872 Gray had cofounded the Western Electric Company in Chicago, which soon became the principal supplier of telegraph equipment to Western Union. Gray was working on a harmonic telegraph at the same time as Bell. At the time of his caveat, Gray did not have a working telephone (neither did Bell), but he did have a different idea for a phone.[16]

According to Ohm's Law, in an electric circuit, voltage equals current times resistance, $V = IR$. Bell's telephone varied the current (I) by changing its magnetic flux in order to carry voice. In his telephone, Gray varied the current instead by changing the resistance (R) of the circuit (sidebar 3.4). In Gray's transmitter, a horizontal diaphragm had a short vertical length of wire suspended from it below. The vibrations from speaking onto the diaphragm caused the wire to dip up and down in a glass of sulfuric acid and water, which varied the resistance of an electric circuit that went through the diluted acid. The circuit carried the disturbance in the current to a receiver with an electromagnet and a diaphragm, which reproduced the original sound.

Bell was aware that variable resistance could also carry sound over an electric circuit, although his telephone design did not use the principle.[17] Bell and Watson built and tested a "liquid" telephone similar to Gray's in March 1876, and it was into this phone that Bell spoke the first words to be carried by telephone: "Mr. Watson—come here—I want to see you."[18] A liquid telephone was obviously impractical and in his Centennial telephone Bell returned to his original idea of varying the current electromagnetically. In a second patent filed in January 1877, however, Bell claimed more broadly as his discovery the use of an undulating current to carry voice and musical tones, not the particular mechanism of transmitting such a current.[19]

In 1875, after learning of the Reis telephone, Thomas Edison had explored the idea of a telephone by varying the resistance. But like other engineers working in telegraphy, he saw telephony as a mere curiosity. In a telegraph caveat filed in January 1876, he claimed a way to transmit sound but did not claim the ability to carry voice. Prompted by Bell's Centennial demonstration, Edison found a more practical way to vary the resistance than Gray's liquid telephone. In 1877 Edison patented a new telephone speaker in which the pressure of speaking onto a diaphragm over a small bed of solid carbon, through which an electric current passed, varied the resistance and carried the tones of the human voice more clearly than Bell's telephone.[20]

Gray and Edison sold their telephone patents to Western Union, which began to produce telephones of its own, prompting the Bell Telephone Company to sue for patent infringement. However, the two companies reached a settlement on November 10, 1879. Telephones at the time could send calls only a distance of about twenty miles and West-

Sidebar 3.4   **The Telephone: Bell and Gray**

Bell's Telephone

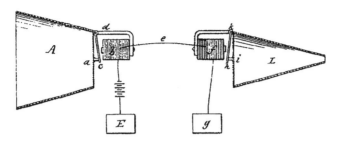

In Alexander Graham Bell's telephone, voice vibrations in cone *A* caused the diaphragm *a* to vibrate armature *c* and the magnetic flux at electromagnet *b*. An electric current connecting the two electromagnets *b* and *f* carried the vibrations to armature *i* and reproduced the sound in cone *L*. The voice in Bell's receiver was, however, faint.

Gray's Telephone

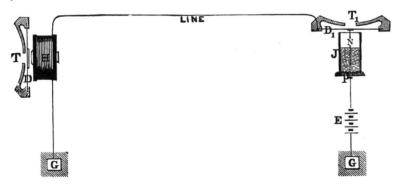

In Elisha Gray's telephone, an electric circuit connected two receivers, *T* and T_1. A voice at T_1 vibrated a diaphragm D_1 and moved a wire *N* up and down in a glass of diluted acid and water. The motion of *N* in the water varied the resistance in the circuit, which varied the current. An electromagnet reproduced the sound on diaphragm *D*.

A telephone requiring a glass of diluted acid was not practical. Edison placed a carbon bed (replaced later by carbon granules) behind the diaphragm in the transmitter, which varied the resistance in the current. The result was a stronger voice in the receiver. Telephones soon employed carbon transmitters.

Sources: for Bell, see Alexander Graham Bell, U.S. Patent No. 174,465 (1876); for Gray, see George B. Prescott, *Bell's Electric Speaking Telephone* (New York: D. Appleton, 1884), p. 15.

ern Union did not see a threat to its established telegraph business. The company was more worried about a threat to its long-distance telegraph monopoly from a group (which soon fell apart) led by the financier Jay Gould. In the settlement, Western Union gave up its claims to the telephone and agreed to share its patents and equipment with Bell Telephone. In return, Western Union received a percentage of the Bell company's receipts over the following seventeen years in which its patent protection would remain in force.[21]

Not wanting to manage a business, Bell resigned from his company's board of directors in 1879 and soon left the company altogether. He and Mabel sold their stock over the next four years to retire as modest millionaires. The other original partners also sold their shares early. As a result, the telephone never produced an owner or group of owners who became as wealthy as Rockefeller, Carnegie, and other industrialists of the period. The high point of Bell's career ended, and he lived comfortably and famously until his death in August 1922; Mabel died five months after him.[22]

Achieving Long Distance

The first telephones connected directly to each other, but such interconnections quickly became impractical as the number of phones in use grew. Instead, phone lines went to a local exchange office, where operators connected subscribers to other phones connected to the same office. Eventually, local exchanges connected to each other through trunk lines. Automatic switching systems appeared in the 1880s, but customers preferred human contact and operators continued to handle most calls (figure 3.6). But the telephone network had to overcome two barriers to its growth, one managerial and the other technical, in order to become a nationwide system.

To manage the new company, Gardiner Hubbard hired Theodore N. Vail (1845–1920), a younger cousin of Alfred Vail (1807–1859), who had helped Samuel F. B. Morse develop the telegraph. Theodore Vail had worked as a railway telegrapher and later managed the railway mail service for the U.S. Post Office. Upon joining the Bell company, he expanded the telephone network by encouraging local phone companies to organize under license, with the Bell firm taking part ownership of each company. In 1881, Vail bought a controlling interest in Western Electric, which manufactured telephone equipment for the Bell System. He also improved transmission

Figure 3.6. Telephone operators in 1895. Source: *Cassier's Magazine*, 8:1 (May 1895): 15.

distances by replacing the steel wires of the network with more conductive copper wires. In 1885 he formed a long-distance subsidiary, American Telephone and Telegraph (AT&T), which absorbed the original Bell company in 1899. However, Vail's drive to invest in capacity rather than reap short-term profits brought him into conflict with new investors, and he resigned in 1887. With the expiration of the second Bell patent in 1894, independent companies began to compete with the Bell System. Telephone ownership surged in the 1890s and the company began to lose its share of the market.[23]

The Bell company realized that it would need to innovate to preserve its still-dominant position; and it did so by overcoming the technical barriers to long-distance calling. Even over copper wires, the strength of calls attenuated from resistance in the lines, with the loss per mile roughly proportional to the resistance. Increasing the cross-sectional area of the copper wire would reduce line resistance but add substantially to the cost of the lines. Wires separated from each other on telephone poles could transmit calls only about one thousand miles, but wires bundled together in underground cables (used mainly in congested areas) had only a fraction of this range.[24]

Figure 3.7. George Campbell. Source: Collected Papers of George Ashley Campbell (New York: American Telephone and Telegraph, 1937), frontispiece. Reprinted by permission of AT&T.

To overcome the limitation on long-distance calls and to improve the reach of cables, AT&T turned to engineers with more advanced scientific and mathematical training. In 1897 the company hired George Campbell (1870–1954), a civil engineering graduate of the Massachusetts Institute of Technology who had taken a master's degree in physics from Harvard and had studied in Europe (figure 3.7). Campbell was aware of the new scientific research in Europe on electromagnetism and wave propagation that led to the development of radio (see chapter 7). He also knew of research that suggested a solution to the problem of long-distance telephony.[25]

A circuit can store electricity and magnetism as well as conduct and resist electric current. Electricity is stored as negative or positive charge and the capacity of a circuit to store electric charge is its *capacitance*. The capacity to store magnetism is its *inductance*. Based on earlier experience with telegraphy, Bell engineers had assumed that attenuation of a phone call was proportional to the product of resistance and capacitance in the line. But the British physicist Oliver Heaviside argued that this rule neglected inductance and that adding inductance to a telephone line would be equivalent to reducing resistance. Following this idea, Campbell placed coils at intervals along a line to add inductance, and experiments found that the added inductance gave long-distance calls greater strength and clarity. Campbell had to place iron cores inside the coils to achieve the desired strength and he had to solve the problem of how far apart to space them. But with these adjustments, the Bell System began to employ inductive loading, as the technique was called, in 1902. The new technique doubled the range of long-distance calls on open wires and tripled it over cables.[26]

A physics professor at Columbia University, Michael Pupin (1858–1935), was working independently on the problem of attenuation, and he patented a version of inductive loading in 1900 just ahead of Campbell. AT&T purchased Pupin's patent rights to keep the innovation away from competitors.[27] Adjusting the electrical properties of the line to compensate for a weak signal was not easy, though, and adding inductance did not make possible full transcontinental telephone calling. A better alternative appeared in 1907, when the radio engineer Lee de Forest invented the triode, an electronic amplifier (see chapter 7). Amplifiers could renew the electrical energy in a call as many times as needed to travel any distance desired. With amplifiers, transcontinental calls finally became practical, as Bell and Watson demonstrated when they emerged from retirement to make the first one in 1915 between New York and San Francisco.[28]

As demand for service grew, the telephone system (like the telegraph network before it) eventually needed a way to send more than one call over the same line at the same time. The telephone industry ironically returned to the principle that Alexander Graham Bell had tried to make practical in his harmonic telegraph, the principle of frequency multiplexing. By 1918 Bell System engineers had found a way to transmit multiple calls over a single line at the same time by giving each call a different electromagnetic frequency range. George Campbell's 1909 invention of the wave filter made it possible to separate these frequencies clearly at each end.[29]

The Bell Legacy

The telephone served a predominantly business clientele in the 1880s and 1890s. The focus of the technology changed, however, as people began to use it for social purposes, and for those who had it, telephone service became a vital part of daily life (figure 3.8). As the cost of service fell in the twentieth century, telephone use spread and gave Americans the means to communicate instantly. In 1907, ownership of AT&T passed from Boston to new investors in New York, and Theodore Vail returned to head the company until 1919. The Bell System became a telecommunications monopoly in 1911 when it acquired its former rival Western Union. When the federal government threatened to break up the Bell System, however, AT&T agreed in 1913 to divest itself of Western Union and allow competing telephone companies access to its long-distance telephone network. The system functioned as a quasi-public utility until its breakup in 1984.[30]

Alexander Graham Bell often called himself a scientist and he associated with prominent scientists in his later life,[31] but Bell was always primarily an engineer. He came to the telephone after joining the search for a better telegraph and he succeeded because those inventors with more experience in telegraph engineering did not recognize the importance of telephony as readily as an outsider like Bell, whose experience lay in human hearing and speech. Electrical engineers in the 1870s were oriented toward the telegraph industry and its incremental needs, which promised the greatest rewards to innovators. Gray himself did not recognize the value of the telephone until Bell was able to commercialize it; and even Edison, in his initial view of the telephone, thought much more as a telegraph engineer than as the visionary he soon proved himself in the field of electric power and lighting.[32] Bell had begun his research in the hope of developing an improved telegraph and his backers had originally demanded that he produce such a device—and not distract himself with telephony. His insight was to see the value of an undulatory current to carry speech.

As the telephone network grew, though, its own needs became more complex. The industry began to need the assistance of engineers with more advanced training in mathematics and science to improve the technology and overcome the obstacles to its expansion. Campbell's adding of inductance to the lines was a clear case of applying a

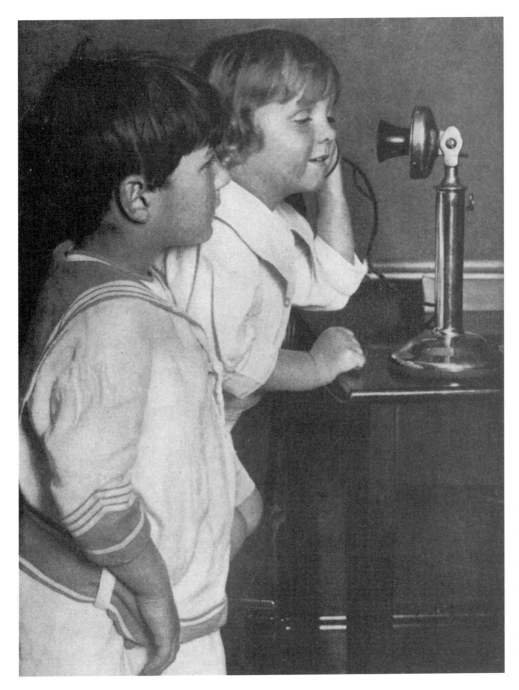

Figure 3.8. King Harris (age three) and his brother Lawrence (age five) in San Francisco speaking with their parents in Washington, DC., in 1916. Source: National Geographic Magazine, 29:3 (March 1916): 301.

scientific principle to meet an engineering need. But inductive loading was not the only scientifically possible answer to the problem of reducing attenuation. Bell engineers could have increased the cross-sectional area of the copper wires, as Edison could have done to reduce power losses in his transmission lines. Like Edison, the Bell engineers chose not to add cross-sectional area to the lines because of the high cost. Inductive loading came into use because scientific constraints left room for engineering choice. Engineering judgment also proved crucial in another key industry at the turn of the century, the chemical process industry of petroleum refining.

CHAPTER FOUR

Burton, Houdry, and the Refining of Oil

The 1870s and 1880s saw the rise of two great chemical process industries in America: the steel industry led by Andrew Carnegie and petroleum refining under John D. Rockefeller. The two industries followed very different paths. After his retirement in 1901, Carnegie sold his company to the banker J. P. Morgan, who merged it with some rivals to form the United States Steel Corporation, bringing 60 percent of the nation's steel-making capacity under the control of one firm. But U.S. Steel faced no serious competitive or regulatory challenges. No other material threatened to take away its market; and the federal government, which tried to regulate or break up other industry-dominant companies, left the steel industry alone. Relying on nineteenth-century processes, U.S. Steel and its rivals did not innovate in any fundamental way between 1901 and 1939. When foreign competitors using more modern methods appeared after 1945, the American steel industry was unprepared and lost ground.[1]

The petroleum industry had to face early challenges. Its continued existence required the timely discovery of new oil reserves. By the turn of the century, the industry also faced the prospect of gradually losing its principal market, the demand for kerosene illuminating oil, as indoor electric light spread. Demand for gasoline to power automobiles gave the petroleum industry a huge new market. But nineteenth-century refining methods could not meet the need for gasoline, and the dominant refiner, Standard Oil, resisted innovation that chemical engineers such as William Burton wanted to make. The near monopoly of Standard Oil aroused public opposition, however, and the U.S. Supreme Court broke up the firm in 1911. The breakup of the company enabled Burton to develop a process that increased the yield of gasoline from a barrel of crude oil so that refining was dramatically less wasteful of petroleum.

The Rise and Fall of Standard Oil

During the nineteenth century, most Americans lived beyond the networks that supplied illuminating gas and later electricity to town dwellers. Until midcentury, Americans relied on candles for indoor light or on lamps that burned whale oil. As the supply of whale oil declined from the killing of too many whales, petroleum-based illuminants began to take its place. Crude petroleum that seeped into ponds and creeks could be refined by simple methods to produce kerosene, a good illuminating oil, and a small industry to refine kerosene arose in the 1850s. The growth of the industry was hampered, though, by the minuscule supply of naturally surfacing crude oil. To supply the market for kerosene, the petroleum industry needed to tap reservoirs of oil that were known to exist underground.[2]

In 1854 a group of Connecticut investors sent a sample of crude petroleum from western Pennsylvania to Benjamin Silliman Jr., a professor of chemistry at Yale University. Silliman reported in 1855 that half of the oil sample could be refined into a high-quality kerosene illuminant and that the rest could have other uses.[3] To look for oil, the Connecticut group hired Edwin L. Drake, a man with no experience in oil or drilling. Drake's previous career as a railroad conductor gave him a free pass to travel, however, and in 1857, with his pass and the title of "Colonel" that he gave himself, Drake went to Titusville in northwestern Pennsylvania. He found an experienced driller, and after nearly two years, the two struck oil from an underground reservoir on August 28, 1859 (figure 4.1).[4]

Drake's example attracted entrepreneurs who soon covered western Pennsylvania in oil wells and simple refining works. Petroleum divides into a range of "fractions" defined by their boiling points (sidebar 4.1). The lighter fuel gases, such as propane, boil at low temperatures. Gasoline boils next, followed by kerosene, diesel and light fuel oils, lubricating oils, and finally the heavy fuel oils and asphaltenes that boil at high temperatures. Early refiners obtained kerosene by heating the crude oil in stills. In a typical still, kerosene vaporized between 400 and 500 degrees Fahrenheit (205 to 260 degrees Celsius) and went into a coiled copper tube, where it condensed into a separate receiving tank. Boiling it again, but at a lower temperature, vaporized out lighter "fractions" in the fluid. Treating the kerosene with an acid and then an alkali removed

Figure 4.1. Edwin Drake (in top hat) at his oil well. Courtesy of Pennsylvania Historical and Museum Commission, Drake Well Museum Collection, Titusville, PA. Mather Photograph No. 4.

Sidebar 4.1   **Petroleum Distillation**

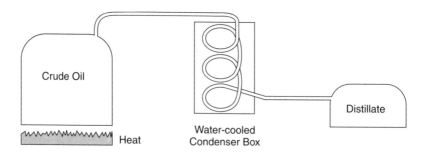

Early oil refining consisted of boiling crude oil. The vapors condensed in a copper tube immersed in running water and formed a distillate. At low temperatures, light fractions evaporated and condensed. Heating the remainder or bottom product at higher temperatures distilled the middle fractions, and further heating distilled the heavier ones. The table gives the broader fractions and their boiling ranges.

Petroleum Fraction	Boiling Range °F	Boiling Range °C
Light naphtha	30–300	−1–150
Heavy naphtha	300–400	150–205
Gasoline	30–355	−1–180
Kerosene	400–500	205–260
Stove oil	400–550	205–290
Light gas oil	500–600	260–315
Heavy gas oil	600–800	315–425
Lubricating oil	>750	>400
Residuum	>1100	>600

Some overlap between fractions occurred. Multiple distillations of the overhead or bottom products were needed to narrow distillates to the desired range.

Sources: James G. Speight, *The Chemistry and Technology of Petroleum* (New York: Marcel Dekker, 1991). Table taken from p. 314 (slightly corrected and edited).

its odor and improved its color, and when the residues of treatment had been removed, the kerosene was ready for sale.[5]

Four years after Drake's well, a new figure entered the growing business of refining oil. Born in upstate New York, John Davison Rockefeller (1839–1937) had moved to Cleveland and left high school two months short of graduation at age fifteen. After working for three years as an assistant bookkeeper in a produce firm, he started his own business shipping produce to the cities of the East Coast. By 1863 he was moderately wealthy and was able to finance an oil refinery in Cleveland. Recognizing the future of the oil industry, Rockefeller left the produce business in 1865 to concentrate on oil refining (figure 4.2).[6]

The high cost of building a steel manufacturing plant kept the number of competing steelmakers small. In oil, the start-up costs of drilling for oil and refining it were much lower, and a large number of small refiners soon competed with each other. By the late 1860s rapid growth in the supply of oil and in refining capacity had outpaced demand, causing prices and profits to fall. The kerosene was also uneven in quality, causing many oil lamps to explode when lit. Rockefeller became the dominant refiner in Cleveland by carefully managing his own refinery and its product and by becoming a reliable supplier to New York and other eastern markets. In 1870 Rockefeller formed the Standard Oil Company to impose a uniform product standard and price.

The railroads suffered from overcapacity themselves in the 1870s, and Rockefeller was able to negotiate preferential rates from them. He invested in new technology to widen his advantage, building his own pipelines from oil fields to railheads and shipping in tank cars rather than in wooden barrels. By purchasing property through veiled subsidiaries, by inviting selected competitors to become his partners, and by driving others out of business through price cutting, Rockefeller brought most of the American refining industry under his control within the span of a decade. In 1873 Standard Oil controlled 10 percent of the refining capacity in the United States; by 1880 the firm had control of 90 percent. Total production of refined oil in the United States grew from a yearly average of 7 million barrels in 1873–75 to an average of 18 million barrels in 1883–85. Four-fifths of this output went to supply kerosene for lamps. Other products, such as solvents and lubricating oils, accounted for the rest of the industry's sales.[7]

Figure 4.2. John D. Rockefeller, circa 1884. Courtesy of Rockefeller Archive Center, Tarrytown, NY.

In the 1880s and 1890s, Rockfeller supplied kerosene to consumers at a price of between five and eight cents per gallon, compared with a kerosene price of about forty cents per gallon in 1870.[8] But Standard Oil's national operation brought it into conflict with state laws and taxes. Some state laws outlawed the ownership of a company in one state by a company outside it, and Pennsylvania taxed the entire firm even if only part of its revenues were earned in the state. To accommodate state laws, reduce tax liability, and veil their degree of industry dominance, Rockefeller and his partners signed the Standard Oil Trust Agreement in 1882. The agreement broke the company into subsidiaries local to each state, and shares in these subsidiaries were held by a trust. As officers of the trust, Rockefeller and his partners retained control, but legally none of the companies owned by the trust owned each other.[9] The Sherman Antitrust Act of 1890 outlawed the Standard Oil trust, which dissolved in 1892, leaving the partners owning the various subsidiaries individually. But New Jersey had passed a law that allowed one company to hold the stock of others, and in 1897 the Standard Oil owners exchanged their shares for those of Standard Oil of New Jersey. Jersey Standard in turn held the stock of the various subsidiaries.

Rockefeller's control of the oil industry had long been controversial and the reconstituted monopoly came under attack after the turn of the century. In a series of magazine articles in 1902–3, Ida Tarbell depicted the company as a threat to democracy.[10] The administration of President Theodore Roosevelt began to enforce antitrust law more vigorously than its predecessors, and in 1906 the United States sued Standard of New Jersey under the Sherman Act. The Supreme Court in 1911 ordered the breakup of Standard Oil into its thirty-three subsidiaries (figure 4.3).[11] A handful of these firms, joined by a small number of new refining companies in Texas and California, still dominated the oil industry. But there were many more refining firms now, and some turned to engineering innovation to gain a competitive advantage.

The Frasch Process and the Need for Gasoline

Despite its dominant market position after 1880, even Standard Oil was far from secure. The company depended on oil reserves in Pennsylvania that were running down; without new supplies, the company faced ruin. In 1885 prospectors discovered vast oil

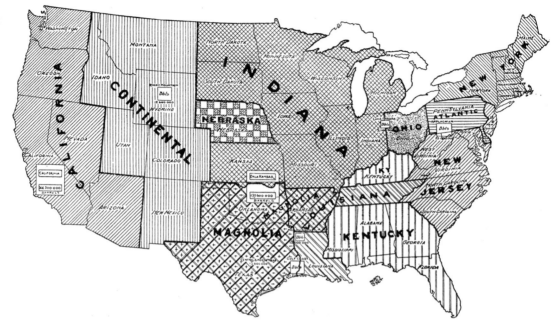

Figure 4.3. Standard Oil companies after the 1911 breakup. Source: Joseph E. Pogue, *The Economics of Petroleum* (New York: John Wiley and Sons, 1921), p. 217.

fields in northwest Ohio and Indiana. But this "Lima-Indiana" oil, as it was called, had a high sulfur content that gave its refined products an unpleasant smell, which existing treatment processes could not eliminate. Rockefeller believed that "the whole future of the Standard depended upon thrusting westward." When his partners objected, he threatened to develop the new oil fields on his own. His partners gave way.[12]

To solve the problems of Lima oil, Rockefeller turned to Herman Frasch (1851–1914), a chemist who worked as a consultant to Standard Oil from 1876 to 1885. Frasch had left the company to buy an oil field near London, Ontario, in Canada whose owner had recently failed because of the oil's sulfur content. After mixing kerosene from the oil with various metallic oxides, Frasch found that reusable copper and lead oxides removed the sulfur. News of his success reached Standard Oil, and Frasch returned to be its chief chemist.[13] After further experimental work, Frasch removed the sulfur from Lima-Indiana oil. Annual U.S. refined oil production rose from 18 million barrels in

1885 to 49 million barrels in 1899, and the new midwestern oil fields accounted for about one-third of this total.[14]

The Frasch process relieved a bottleneck in the supply of illuminating oil, but it could not save the industry from a much worse fate, shrinking demand. Although rural areas continued to buy kerosene illuminants until electrification arrived later in the twentieth century, electric power began to replace oil as a lighting source in urban and suburban America in the 1880s and 1890s. Petroleum refining would have gone the way of the gaslight industry if the automobile, with its gasoline-burning engine, had not created a new and even greater demand for oil after 1900. The promise of a new market brought with it a new challenge.

Simple or "straight-run" distillation methods, in which crude oil was boiled and condensed, yielded about 30 to 50 percent kerosene from a batch of crude oil. But no more than about 20 percent of crude oil could be distilled into gasoline.[15] To meet the need for more gasoline, drillers began to search for new supplies of crude oil, but refining only a fraction of new oil was wasteful. Early automobiles were luxuries that did not stretch gasoline supplies at first, but Henry Ford's introduction in 1908 of a mass-produced affordable car, the Model T, caused demand for gasoline to soar. The oil industry needed a more efficient way to supply this market.

William Burton and Thermal Cracking

William M. Burton (1865–1954) found a way to obtain more gasoline from a given amount of crude oil (figure 4.4). After earning one of the first American Ph.D.s in chemistry, from Johns Hopkins University in 1889, he went to work as Frasch's assistant and then moved in 1890 to a new refinery at Whiting's Crossing, Indiana, seventeen miles east of Chicago. Burton did not feel challenged by routine laboratory work, though, and he moved into management, becoming a vice president of Standard Oil of Indiana, the principal midwestern subsidiary of the Standard Oil group, in 1903. The Whiting refinery reported to him, and his attention returned to research as the need for more gasoline became urgent after 1908. Dr. Robert E. Humphreys had taken charge of the laboratory at Whiting, and at Burton's direction, Humphreys and an assistant, Dr. F. M. Rogers, began to study ways to obtain more gasoline from crude oil (figure 4.5).[16]

Figure 4.4. William M. Burton.
Courtesy of BP America, Inc.,
Chicago, IL.

Crude petroleum consists mostly of hydrocarbons, molecules composed of hydrogen and carbon atoms. Burton and Humphreys knew that heavier molecules could break apart into lighter ones. They decided to see if they could get more gasoline out of a barrel of crude oil by breaking or "cracking" heavier petroleum molecules into lighter ones in the gasoline range (sidebar 4.2).[17] Under Burton's direction, Humphreys at first tried different ways of heating the heavier oil to obtain more of the lighter vapors, with poor results. Next, he tried mixing in aluminum chloride, which was known to break heavier oil molecules into lighter ones. This approach used too much oil, though, and the aluminum chloride was expensive and could not be reused. As Burton later wrote, "our efforts were not successful . . . we met failure in every direction."[18]

There remained one last possibility. Simple distillation subjected crude oil to varying temperatures but not to varying pressures. "It had been known for a long time that distillation of petroleum products under pressure resulted in their disassociation and

Figure 4.5. Robert Humphreys in the Whiting laboratory, circa 1908. Courtesy of BP America, Inc., Chicago, IL.

· ·

Sidebar 4.2   **Petroleum and Chemical Reactions**

Gasoline

Petroleum consists of molecules composed mostly of hydrogen (H) and carbon (C) atoms. Gasoline molecules boil between 30° and 355° F and divide into sub-fractions that can have between 4 and 12 carbon atoms each. A sub-fraction called pentane, for example, consists of molecules having five carbon atoms. Hexane has six carbon atoms, heptane seven, and octane eight.

The number of hydrogen atoms in gasoline also varies. Molecules belonging to the *paraffin* or *alkane* group have twice the number of hydrogen atoms, plus two, as carbon atoms (C_nH_{2n+2}). Octane (C_8H_{18}), a paraffin, has eight carbon atoms and eighteen hydrogen atoms. Molecules in the *olefin* or *alkene* group have only twice the number of hydrogen atoms as carbon atoms (C_nH_{2n}). Ethylene (C_4H_8) is an olefin.[1]

Chemical Reactions

Molecules can combine with each other or break apart in chemical reactions, forming new molecules. The *first law of thermodynamics* states that when substances undergo these changes, the total mass (the number of atoms) remains the same. The atoms redistribute themselves. In the reaction (indicated by the arrow)

$$2C_{14}H_{30} \;\rightarrow\; C_8H_{18} \;+\; C_{20}H_{42}$$

tetradecane   octane   eicosane

two molecules of tetradecane ($2C_{14}H_{30}$), a kerosene, turn into one molecule of a gasoline, octane (C_8H_{18}), and one molecule of a fuel oil ($C_{20}H_{42}$), eicosane. The total number of carbon (28) and hydrogen (60) atoms remains the same before and after the reaction, but the two original molecules change into two different ones. The breaking of the two kerosene molecules is known as *cracking*.

[1] Crude petroleum contains thousands of known and many as yet unidentified molecules, and new molecules are synthesized by the refining process. Fractions are defined by the preponderant molecules in each fraction.

Source: James G. Speight, *The Chemistry and Technology of Petroleum* (New York: Marcel Dekker, 1991), pp. 219–32, 473–77.

· ·

production of some low-boiling and some high-boiling fractions," recalled Burton, "but this process had never been applied in a practical way . . . owing to the extreme hazard."[19] Under high heat and pressure, crude oil could explode in a terrific fireball. But there seemed no other way to get more gasoline than to take this risk. Humphreys conducted experiments in which he heated diesel oil in a fifty-gallon steel tank under rising pressures as well as rising temperatures. A condenser received the vapors and took them to a receiving tank. For safety, Humphreys raised the pressures gradually. Finally, at a pressure of seventy-five pounds per square inch (about five times atmospheric pressure) and a temperature of about 700 degrees Fahrenheit, the still produced gasoline of acceptable quality. The process doubled the amount of gasoline that a barrel of crude oil could produce, from about 20 to about 40 percent.[20]

Burton asked Standard Oil's headquarters in New York for $1 million to build full-scale pressure stills. After Rockefeller's retirement in 1897, however, the monopoly had come under more conservative management that refused to consider such a dangerous-sounding scheme. One of the Standard Oil directors in New York was reported to have said: "Burton wants to blow the whole state of Indiana into Lake Michigan."[21] The breakup of Standard Oil in 1911 came just in time for Burton and his team. The Whiting refinery went to the newly independent Standard Oil Company of Indiana (later known as Amoco, now BP America). The new company had refining capacity but no oil reserves, and its leaders threw their support behind Burton's work. On July 3, 1912, he filed a patent on the new process, assigning the rights to the company. The patent was granted on January 7, 1913.

The Burton Patent

The innovation in the Burton process was its mechanical arrangement of the boiler and condenser (sidebar 4.3). To heat crude oil under pressure, Burton at first placed a shut-off valve between the boiling tank and the condenser, so that all of the heat and pressure would be confined to the boiling tank. But the condensed vapors produced in this way had too many unwanted by-products and an unacceptable odor. Humphreys moved the shut-off valve to a different location, between the condensing tube and the final receiving tank. Heating the heavier fractions of oil then created pressure in both the boiler and the condenser, not just in the boiler. The effect of making this simple move was dramatic: heavier oil cracked into acceptable gasoline. In addition to breaking heavier molecules, the cracking also produced molecules lighter than gasoline. Some of these combined with each other to form gasoline molecules. The result was a substantial increase in the amount of gasoline produced from a barrel of crude oil.[22]

Sidebar 4.3 **Thermal Pressure Cracking**

The Burton Process

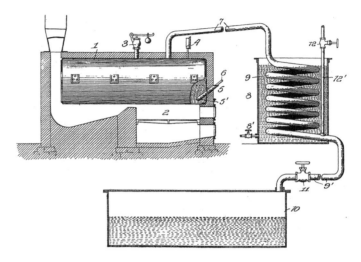

The Burton patent outlined a mechanism of boiler (*1*), condenser (*9*), receiving tank (*10*), and shut-off valve (*11*). Instead of placing the shut-off valve between the boiler and condenser, the design placed the valve between the condenser and the receiving tank. Heating crude oil in the boiler and condenser at about five atmospheres of pressure and 700° F cracked heavier kerosene into lighter gasoline.

Cracking Heavier Molecules into Gasoline

The cracking process could produce new molecules that were heavier as well as lighter than gasoline. In the example from sidebar 4.2, two molecules of tetradecane could be cracked into one of octane and one of eicosane, a heavier fuel oil:

$$2C_{14}H_{30} \rightarrow C_8H_{18} + C_{20}H_{42}$$

tetradecane   octane   eicosane

(kerosene)   (gasoline) (fuel oil)

Cracking the eicosane added a new octane molecule and left a new one of kerosene, dodecene:

$$C_{20}H_{42} \rightarrow C_8H_{18} + C_{12}H_{24}$$

eicosane   octane   dodecene

Source: William Burton, U.S. Patent No. 1,049,667 (1913).

The two engineers did not achieve this breakthrough by applying a scientific theory that predicted the results they would find. As Burton stunningly observed in his patent, the results occurred for chemical reasons "which I do not attempt to explain." In an address to the Chicago section of the American Chemical Society, at which he was awarded the Josiah Willard Gibbs medal in 1918, Burton credited science with a role in the oil industry that began with Frasch and his process: "This was the starting point for a better feeling between the chemical profession and the petroleum industry, and from that time, more and more chemists have been employed in the refining industry, until to-day the larger refineries depend almost entirely upon chemists to manage, not only the refinery as a whole, but the various departments of the same."[23] But, in fact, what Burton attributed to science was a new kind of engineering, chemical engineering.

chemical engineering

The aeronautical historian Walter Vincenti has described a way of understanding flows, called _control-volume analysis_, that sheds light on what Burton and his team achieved. Developed in the nineteenth century, control-volume analysis in essence held that if a flow going into a box was known and the flow going out was also known, it did not matter to engineering exactly what went on inside the box. To scientists, in contrast, a proper understanding required knowing in a precise way what occurred inside the box. As Vincenti observed of fluids in motion,

> All nontrivial flows . . . vary either in space or time (or both), and a physicist in his quest for knowledge characteristically wants to know the point-by-point details of this variation. For this end the overall results of control-volume analysis are not enough. In fluid mechanics physicists therefore concentrate on solution of the differential equations of motion [i.e., equations that reduce matter to points and calculate how the points move in space and time]. . . .
>
> In engineering the situation is very different. Engineers, unlike physicists, are after useful artifacts and must predict the performance of the objects they design. . . . Moreover, the problems that arise often present serious difficulties in the underlying physics or in the solution of differential equations, so that more than overall results are not feasible.[24]

Burton did not refer in his patent to control-volume analysis but his innovation was a textbook example of it. The thousands of simultaneous chemical reactions that took place in the thermal cracking of petroleum would have been impossible to isolate and calculate individually. Burton controlled what went into the system, he knew what came out, and he did not try to explain the chemical reactions that occurred along the way. But his approach was not a matter of cutting corners. It expressed the wisdom of knowing the limits to sophisticated analysis as well as of knowing when detailed and exact understanding was necessary and proper.

Building and operating full-scale pressure stills challenged Burton and his colleagues yet again. When Burton asked mechanical engineers for help building a containment vessel, he noted drily, "we did not receive very much encouragement."[25] The steel plates had to be riveted together, and their strength was uncertain at high pressures and temperatures. Thermal cracking required great courage on the part of refinery workers, who had to plug flaming leaks. These dangers lessened with the development of electric welding in the late 1920s and other improvements in still design.

Standard Oil of Indiana began producing cracked gasoline in January 1913 (figure 4.6), at about the time that Henry Ford perfected his method of mass-producing automobiles. In 1899 the United States produced about 6 million barrels of gasoline, for use mainly as cleaning solvents. By 1919, mostly to supply motor vehicles, U.S. gasoline production had climbed to 99.7 million barrels, of which 15.5 million (16 percent) consisted of cracked gasoline. By 1929 U.S. gasoline production reached 441.8 million barrels, of which 143.7 million (33 percent) was cracked gasoline (figure 4.7).[26]

Eugene Houdry and Catalytic Cracking

The Burton process was limited in one important way: it produced gasoline only in batches and had to stop approximately every two days so that the stills could be cleaned of carbon residue that built up during the refining process. Engineers working for the Texas Company (now Texaco) and for Standard Oil of New Jersey (later Exxon, now ExxonMobil) soon devised thermal cracking processes that could go for longer periods between cleanings. The most notable of these was a process invented by Jesse Dubbs and his son, whom he gave the prophetic name Carbon Petroleum Dubbs. Patented between 1915 and 1919, the Dubbs process could feed crude oil in and out of pressure stills on a nearly continuous basis, removing carbon residue as it did so. Thermal

Figure 4.6. Burton stills in operation. Courtesy of BP America, Inc., Chicago, IL.

cracking and processes to improve it soon became entangled in patent litigation until the major refining companies agreed to share their processes in 1931.[27]

During the early 1920s, however, the petroleum refining industry faced a new technical challenge. The gasoline produced for motor vehicles suffered from a problem called engine knock, in which the gasoline failed to ignite evenly. As automobiles developed stronger engines to drive faster on newly paved roads, this problem became more severe. In the early 1920s, Charles Kettering and Thomas Midgley of General Motors found that adding tetraethyl lead to gasoline could sharply reduce engine knock. Motor vehicles ran on leaded gasoline from the 1920s until the 1970s, when

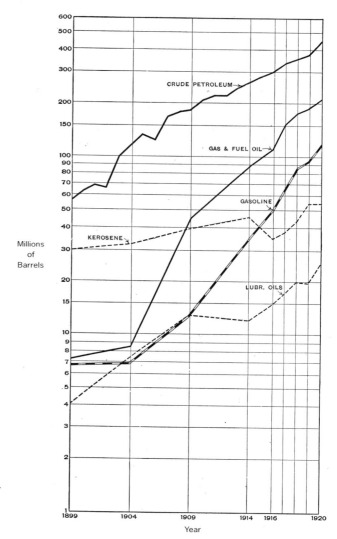

Figure 4.7. Trend in output of crude oil, 1899–1920. Source: Joseph E. Pogue, *The Economics of Petroleum* (New York: John Wiley and Sons, 1921), p. 84.

safety concerns over air pollution brought new emission standards that removed lead from gasolines used by newer vehicles.[28]

To measure the antiknock properties of gasoline, Graham Edgar of the Ethyl Corporation devised a scale in 1926. Edgar rated gasoline with a number between 0 and 100 according to how its performance compared with two benchmark gasolines, normal heptane (rated 0, severe knocking) and iso-octane (rated 100, no detectable knocking).

The average of two different tests gave numbers on this scale called the *octane rating* of gasoline, with an average closer to 100 being better.[29] By 1930 gasolines reached octane ratings between 60 and 70. But the best thermal cracking could not raise the octane above 82. Newer and more powerful automobile engines required higher-octane gasoline, and aircraft engines needed fuel with octane ratings of nearly 100.[30] The design of an efficient catalytic cracking process by the French engineer Eugene Houdry produced higher-octane gasoline and also increased the supply (figure 4.8).

Born in France and trained as a mechanical engineer, Eugene J. Houdry (1892–1962) joined his father's steel company after serving in the First World War (1914–18). To relieve the French need for imported oil, Houdry tried in the early 1920s to convert domestic coal into oil. This proved uneconomical. But his other passion was automobile racing, and he owned and raced a Bugatti. On a visit to the United States in 1922, he attended an Indianapolis 500 race as a spectator and then visited the Ford plant in Detroit. Houdry realized that the limitation on speed was not just in the mechanical design of cars but in the quality of the gasoline.[31]

During the nineteenth century, chemists had discovered that adding certain chemicals, called catalysts, to a mixture accelerated chemical reactions. Burton and Humphreys at first tried to use a catalyst, aluminum chloride, to crack petroleum; however, the material was uneconomical, and the two shifted to thermal cracking. Houdry knew that cheaper catalysts released carbon by-products that stuck to the catalyst and interfered with its action. He believed that finding a catalyst that was economical and easy to clean would be the next great advance in the making of gasoline. Like Edison in his search for a proper filament, Houdry tried hundreds of possible catalysts. Finally, in 1927, Houdry found that fluid petroleum molecules adhered to solid oxides of silicon and aluminum (in pellet form) and that the catalysts cracked heavier petroleum molecules into lighter ones. The process required higher temperatures but lower pressures than thermal cracking. A carbon residue accumulated on the catalysts with repeated use, but by regularly burning the catalysts with air and heat, the residue could be removed. Houdry proved that catalytically cracked gasoline was as good as the best gasoline on the market by using it to accelerate his Bugatti racecar up a steep hill.[32]

Unable to obtain funding in France, Houdry interested the Vacuum Oil Company of New Jersey in his process. In 1930 he set up a pilot plant at the Vacuum research

Figure 4.8. Eugene J. Houdry. Source: Harold F. Williamson *et al.*, *The American Petroleum Industry: The Age of Energy 1899–1959* (Evanston, IL: Northwestern University Press, 1963), p. 613. Courtesy of Air Products Corporation.

center in Paulsboro, New Jersey. The acquisition of Vacuum by Standard Oil of New York (known as Socony, later Mobil, now part of ExxonMobil) stalled Houdry's research, but Sun Oil of Philadelphia (Sunoco) stepped in to carry it on. In 1936, backed by Socony-Vacuum and Sun, Houdry began producing gasoline for sale. In the best thermal cracking processes, a charge of heavier oil could yield about 40 percent gasoline, with octane ratings between 72 and 82. With catalytic cracking, the same amount of heavier oil produced almost 60 percent gasoline, and if the process was repeated, the amount reached 75 percent. The octane ratings on the first "pass" were between 81 and 95 and on the second between 83 and 95.[33]

Catalytic cracking increased the octane rating mainly by *rebranching* atoms. Gasoline molecules with the same numbers of carbon and hydrogen atoms can differ in the way the atoms link together (sidebar 4.4). Gasoline molecules with their atoms arranged in *straight chains* have low octane ratings because they tend to break and combust prematurely, while molecules with atoms in *branched chains* performed much better. Catalytic cracking greatly increased the proportion of branched molecules in gasoline.[34]

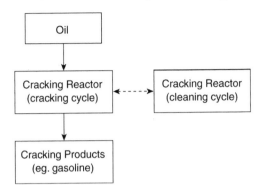

Sidebar 4.4   **Catalytic Cracking**

In the Houdry process, oil reacted with catalysts in a reactor vessel from which gasoline was tapped. The catalysts were then "regenerated" (cleaned by burning) in the reactor and reused. Reactors alternated between cracking and cleaning. In later moving-bed processes (not shown), the catalysts went to a separate regenerating chamber. The cleaned catalysts were returned for reuse.

Catalytic cracking takes advantage of the ability of atoms in gasoline molecules to bond with each other in two kinds of chains:

Straight chain (normal pentane C_5H_{12})   Branched chain (isopentane C_5H_{12})

In *straight chains*, the carbon atoms bond in a single line, with hydrogen atoms surrounding and linked to them. Paraffin molecules with straight chains are called "normal." Molecules can also take the form of a *branched chain*, indicated by the prefix "iso-", with the carbon atoms branched instead of arranged in a single line. Branched molecules are not as easily broken in an internal-combustion engine as straight-chain molecules and thus are less prone to cause engine knock by igniting prematurely.

In the octane rating scale, normal heptane (benchmark zero) is a straight-chain molecule and iso-octane (benchmark 100) is a branched one. Catalytic cracking produces more gasoline with branched-chain molecules and thus higher octane ratings.

Source: Daniel DeCroocq, *Catalytic Cracking of Heavy Petroleum Fractions* (Paris: Editions Technip; Houston TX: Gulf Publishing (distributor), 1984), pp. 73–80.

Houdry's process of passing fuel oil through solid pellets of silica-alumina came to be called *fixed-bed* catalytic cracking. In the *moving-bed* processes developed a few years later, the catalyst consisted of a finely powdered catalytic material. This material provided greater surface area to which gasoline molecules could adhere and crack, and its use permitted a continuous process. The finer catalysts could flow out of the reactor with the oil. After separation, the catalysts were burned ("regenerated") to remove carbon desposits and then returned to the reactor for reuse. After 1945 moving-bed catalytic cracking for the most part replaced the fixed-bed process, and more efficient synthetic catalysts later replaced natural ones. Engineers in the 1930s also learned how to produce gasoline in other ways. In a process called *alkylation*, lighter fractions of oil were combined to create gasoline. In *reforming*, gasoline molecules of the same size were rebranched to give higher-octane performance.[35]

Burton and Houdry worked within a framework of basic chemical ideas provided by science. But their designs were not predicted by these ideas, and if they had tried to follow an approach that sought an exact understanding in the manner of physics, their efforts to innovate would have failed. They took a broad view of engineering as well and looked beyond the specialized fields in which they had been trained. Burton, a chemist, created an innovation that was essentially mechanical; Houdry, a mechanical engineer, found the chemicals that made catalysis in petroleum refining practical. The modern oil refinery needed mechanical engineering to work and required the twentieth century's greatest machine, the automobile, to provide a market for what it produced.

CHAPTER FIVE

Ford, Sloan, and the Automobile

*T*he automobile is the preeminent machine of the modern world. In 1900 there were 8,000 registered automobiles in the United States; by 1939 there were 26 million.[1] The car gave Americans a personal mobility and freedom unknown in the nineteenth century, and by 1939 auto manufacturing had become America's leading industry. But in 1900 the future of the car was hardly clear. The market for automobiles then was small and exclusive, and cars powered by steam and electricity vied with those powered by gasoline. The car began to reach a mass market after Henry Ford's introduction of the gasoline-powered Model T in 1908. America eventually outgrew Ford's vision of a rugged egalitarian car, and a great rival, the General Motors Corporation under Alfred P. Sloan Jr., pulled ahead of Ford in the 1920s by meeting a growing demand for variety.

The Coming of Automobiles

The railway locomotives of the nineteenth century were external-combustion engines, in which a coal-fueled boiler heated water into steam. The steam went at high pressure into side cylinders, where it pushed pistons and rods attached to them. These rods turned the locomotive wheels. At the 1876 Philadelphia Centennial Fair, Niklaus Otto of Germany demonstrated a new kind of piston engine in which the fuel burned inside the piston cylinder and not in a separate compartment (figure 5.1). Otto's internal-combustion engine went through a cycle of four piston strokes. On the first downstroke, a mixture of coal gas and air entered the piston cylinder. On the upstroke, the piston head compressed the mixture. An electric spark then ignited the fuel, causing another downstroke that delivered power to turn an attached wheel. A second upstroke exhausted waste gases.[2] Modern automobile engines adopted the four-stroke cycle

internal combustion

79

Figure 5.1. Otto Engine at the 1876 Philadelphia Centennial Fair. Source: Phillip T. Sandhurst, *The Great Centennial Exhibition* (Philadelphia: P. W. Ziegler, 1876), p. 358.

Benz
Daimler

(sidebar 5.1). Otto sold his machine as a stationary engine. In 1885 the German Karl Benz placed a gasoline-fueled engine on a three-wheeled carriage and began making cars. Gottlieb Daimler, whose company later merged with Benz, built a more stable four-wheeled gasoline car the following year, and Emile Levassor of France built a car with the engine mounted in front instead of under the seat or in back. The brothers Charles and Frank Duryea of Springfield, Massachusetts, designed and built a four-wheeled gasoline car in the United States in 1893. Americans soon began to drive the new vehicles (figure 5.2).[3]

The early gasoline car was not an easy machine to own. The engine was noisy, produced a sooty exhaust, and needed highly flammable fuel. A driver had to turn a hand

Sidebar 5.1   **The Internal-Combustion Engine**

External and Internal Combustion

A railway locomotive employed an external-combustion engine that burned fuel in a separate compartment to heat water into steam. The steam heated under pressure and then went to piston cylinders on each side of the locomotive. The high-pressure steam drove pistons connected to rods in a reciprocating motion that turned the locomotive wheels.

In an internal-combustion engine, fuel burned directly in a piston cylinder in a series of timed bursts. The engine drew a mixture of fuel and air into the cylinder and then compressed and ignited it. The combustion pressure drove the piston, which in the Otto engine turned a wheel. Otto's engine was designed for stationary use in factories and burned a mixture of coal gas and air. In motor vehicles, liquid fuel was more efficient, and cars burned a mixture of gasoline and air.

The Four-Stroke Engine Cycle

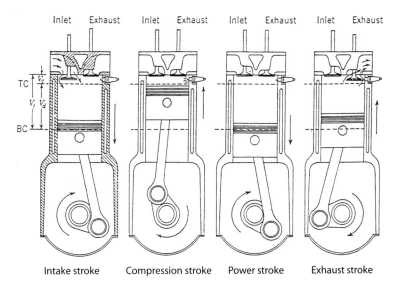

In modern automobile engines (above), the piston goes through the Otto cycle of four strokes. An intake stroke draws a mixture of fuel and air into the cylinder. A compression stroke compresses the mixture for ignition. A spark then ignites the mixture, causing a combustion or power stroke. An exhaust stroke expels the waste gases.

Source: Lynwood Bryant, "The Origin of the Automobile Engine," *Scientific American* 216, no. 3 (March 1967): 102–12. Cylinder diagrams from John B. Heywood, *Internal Combustion Engine Fundamentals* (New York: McGraw-Hill, 1988), p. 10.

Figure 5.2. Jane Newkirk (grandmother of senior author) in an early car. Source: Billington family album.

steam and electric cars

crank in front of the car to start moving the pistons, and mechanical breakdowns were frequent. Steam and electric vehicles at first offered attractive alternatives. Steam engines had been used to power carriages and tractors in the nineteenth century. In 1898 the Stanley brothers began manufacturing steam-powered cars in Watertown, Massachusetts, that burned kerosene instead of coal. Stanley steamers could drive as fast as gasoline cars with internal-combustion engines. The Stanley cars took thirty minutes to start and had to stop frequently for water, but after 1901 steamers made by the rival White firm began to use preheated water and condensers to recycle steam. The Stanleys eventually adopted similar devices. Steam cars were complex and difficult to maintain, though, and most steam car makers switched to manufacturing gasoline cars with internal-combustion engines in the decade after 1900. The Stanleys continued to make steam cars but preferred to make them by hand, which limited their market to several thousand owners. The Stanley firm went out of business in the mid-1920s.[4]

Electric battery-powered cars also appeared in the 1890s. These could be recharged with direct current at central power stations, which were happy to sell power in hours of off-peak usage. With rectifiers to convert A.C. to D.C., alternating current could be used for recharging as well. Because electrics were silent, clean, easy to drive, and reliable, they found a ready market, especially among women. Electric cars had a range of about fifty miles and needed several hours to recharge, but they were well-suited for driving short distances in town. Electric cars failed, though, for cultural reasons as much as for technical ones. Gasoline cars appealed to men, who enjoyed the challenge of driving and maintaining them. As gasoline cars gradually increased their speed, a key attraction to male drivers, they left electrics behind. Touring the countryside by automobile was also an ambition of most early car buyers and the limited range between rechargings meant that electrics could not be used for travel over long distances. Women who had bought electric cars remained loyal to them; but Charles Kettering's 1911 invention of an electric self-starter made hand cranking unnecessary, and new women drivers began to switch to gasoline cars. Canned gasoline was easy to distribute through the network for distributing kerosene illuminants, and repair garages later sold gasoline from pumps. The electric power network did not reach most suburban households until the 1910s and 1920s, too late to give electric cars outside the cities a nearby supply of electricity.[5]

But in one respect cars using internal-combustion engines were no better than early steam and electric vehicles: their high cost. Purchase prices of $500 to $1,000 or more at the turn of the century permitted only the wealthy to buy automobiles. In Europe, cars were built in small numbers by firms whose names became synonymous with wealthy owners, such as Rolls-Royce in England and Bugatti in Italy. In the United States, luxury cars like the Pierce-Arrow also served a small market that could afford them.[6] The manufacturing of early cars followed craft traditions, in which a few highly skilled workmen assembled each car in place in small workshops or factories. Buyers usually purchased cars directly from the factory and relied on urban garages or their own servants to maintain and repair them. The small scale of early automobile making also made it possible, however, for entrepreneurs with little capital to enter the business. Henry Ford was one of these entrepreneurs.

electric cars

83

Henry Ford

Henry Ford (1863–1947) grew up on a farm outside Detroit, Michigan (figure 5.3). He was fascinated by engines at an early age, worked as a steam engine repairman in the 1880s, and studied an Otto engine that came to the area. In 1891 he left for Detroit, where he worked as an engineer for the city's Edison Illuminating Company and built a gasoline car in his free time. In June 1896 he successfully drove the car, a motorized carriage on four bicycle wheels that he called a "quadricycle." In August, at a convention of the Edison companies in New York City, Ford met Thomas Edison, who encouraged him to keep working on a gasoline car. With the backing of the mayor and other investors, Ford resigned from Edison in 1899 and formed the Detroit Automobile Company. His share in the company was meagre, though, and his failure to produce a workable design forced the company to dissolve in November 1900.[7]

To raise capital for a new company, Ford decided to win fame as a racecar builder. Ever since the steamboat, new forms of transportation had proved themselves in races, either to beat a fixed time or to race competitors to finish in the fastest time. In 1807 Robert Fulton's steamboat reached a speed of four miles per hour in a trip up the Hudson River from New York City to Albany and back. In 1896 Frank Duryea beat a Benz car in a road race from New York City to Irvington-on-Hudson and back.[8] Returning to the family farm, Henry Ford built a racecar and entered a meet held at Grosse Pointe, Michigan, on October 10, 1901, with a prize of $1,000. Alexander Winton of Cleveland, a carmaker and the country's foremost racecar driver, also entered and was expected to win.

After the starting gun, Winton took an early lead but Ford began to gain on him as the local crowd cheered him on. Winton's car then developed engine trouble and Ford caught up and won the race, gaining him the gratitude of Detroit and the backing of new investors to form a second company. The investors didn't give Ford the freedom he wanted, though, and he left in March 1902. The company reorganized under Henry Leland and took the name Cadillac. Ford gave racing one last chance. On October 25, 1902, Winton returned to Grosse Pointe to avenge his earlier defeat, and Detroit eagerly awaited the rematch. A racing cyclist, Barney Oldfield, agreed to drive a new and more powerful car that Ford built (figure 5.4). Winton did not finish the race and Ford's car won, setting an American record of under one minute six seconds per mile.[9]

Figure 5.3. Henry Ford, circa 1904. Courtesy of The Henry Ford Museum, Dearborn, MI. Photograph no. P.O. 1282.

Ford organized a third company, the <u>Ford Motor Company,</u> in June 1903 with the principal backing of Alexander Malcomson, a coal dealer. Ford had a larger share now, and he ran the engineering side while Malcomson's accountant, James Couzens, managed the rest of the business. The new firm survived and expanded by innovating in two ways. First, with the help of capable machinists such as C. Harold Wills, Ford introduced cars that embodied his intuition, against the conventional wisdom of the time, that cars could be lighter in weight and greater in power. Second, Couzens established a strong network of dealerships. He demanded that dealers pay the company in advance for the cars they sold instead of paying after selling them. Distributors also had to offer repair

85

Figure 5.4. Henry Ford and Barney Oldfield next to Ford's 999 racer. Courtesy of The Henry Ford Museum, Dearborn, MI. Photograph no. P. 188.4568.

services to owners. Dealers who accepted these conditions discovered that they could make money, and the guarantee of service helped build the reputation of Ford cars.[10]

Ford's first car, the Model A, sold 1,700 cars in its first fifteen months. With this success, Ford was able to stay in business without having to rely on outside money. But his desire to make low-priced vehicles for a larger market came into conflict with Malcomson and his partners, who wanted the company to produce higher-priced models for wealthier buyers. After producing two high-priced models, the Models B and K, Ford bought out Malcomson in July 1906 and became majority owner of the company. He

Figure 5.5. A 1908 Ford Model T. Courtesy of The Henry Ford Museum, Dearborn, MI. Photograph no. A-4810.

quickly brought out a new car, the Model N, that weighed 1,050 pounds but could reach a speed of forty-five miles per hour and sold for $600. The Model N raised Ford Motor Company sales from 1,600 cars in the 1905–6 season to 8,243 in 1906–7.[11] Ford set his sights even higher and began planning the breakthrough car that became the Model T.

The Revolution of 1908

Ford rolled out the first Model T in 1908 (figure 5.5). The new car weighed 1,200 pounds and could go no faster than the Model N, but it delivered greater power to the wheels and incorporated several important innovations, principally the use of vana-

3 core systems

dium steel, a lightweight alloy that afforded the strength of a heavier car. Ford also designed the car for unpaved country roads with the body high above the ground. The Model T had an ungainly appearance as a result but could drive through mud and grass that other cars could not. Shifting gears was simpler than in any other car on the market, a crucial attraction to the ordinary buyers Ford wanted to reach. He also designed the car so that it could be understood and maintained by its owner.[12]

The Model T depended on three core systems to work: electric ignition, chemical combustion of the fuel, and mechanical transmission of power from the engine to the wheels. The Model T engine had four vertical cylinders and a piston in each cylinder connected to a single crankshaft underneath. When the pistons began moving, the crankshaft rotated. Gears (in a box) transmitted this rotation to a central driveshaft running lengthwise under the car to gears that turned the rear axle and its wheels (figure 5.6).[13]

To start the car, a driver turned a hand crank in front. The crank turned the crankshaft and also turned a magneto, a circular case in which a flywheel with magnets rotated around a fixed wire coil. The flywheel generated electricity in the same manner as a rotating armature in an electric generator. The electricity went to four small coil units (transformers) that stepped down the current and raised the voltage. The voltage from each coil went through a revolving contact called the distributor that fired a spark plug in each cylinder in a rapid sequence (sidebar 5.2). The firing ignited a mixture of gasoline and air that had entered the cylinder through a carburetor, and the combustion pushed down the piston in each cylinder.

The crankshaft connected the pistons so that when two pistons went down, the other two went up. Each cylinder went through a cycle of four strokes. Engineers measured the average power of combustion on the piston head using the same formula, *PLAN/* 33,000, that James Watt had developed in the eighteenth century for measuring the horsepower in a piston steam engine. In an internal-combustion engine, automotive engineers called the result of this formula the *indicated horsepower* (sidebar 5.3).[14] In the Model T at its top speed of thirty-seven miles per hour, the indicated horsepower was about twenty-three. Friction losses in the cylinders caused the power that reached the crankshaft, known as the *brake horsepower*, to be less, just under twenty. Further losses in the transmission of brake horsepower to the wheels reduced the power at the wheels, or *traction horsepower*, to about twelve. The Model T was still an efficient car for its time.[15]

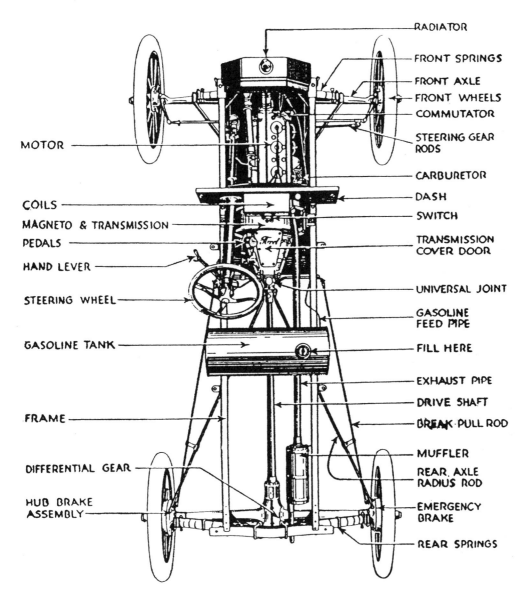

RADIATOR

FRONT SPRINGS

FRONT AXLE

FRONT WHEELS

COMMUTATOR

STEERING GEAR
RODS

CARBURETOR

DASH

SWITCH

TRANSMISSION
COVER DOOR

UNIVERSAL JOINT

GASOLINE
FEED PIPE

FILL HERE

EXHAUST PIPE

DRIVE SHAFT

BREAK PULL ROD

MUFFLER

REAR AXLE
RADIUS ROD

EMERGENCY
BRAKE

REAR SPRINGS

MOTOR

COILS

MAGNETO & TRANSMISSION

PEDALS

HAND LEVER

STEERING WHEEL

GASOLINE TANK

FRAME

DIFFERENTIAL GEAR

HUB BRAKE
ASSEMBLY

Figure 5.6. Model T chassis. Source: *Ford Manual: For Owners and Operators of Ford Cars* (Detroit: Ford Motor Company, 1914), p. 4.

Sidebar 5.2 **The Model T: Ignition and Fuel**

Electric Ignition

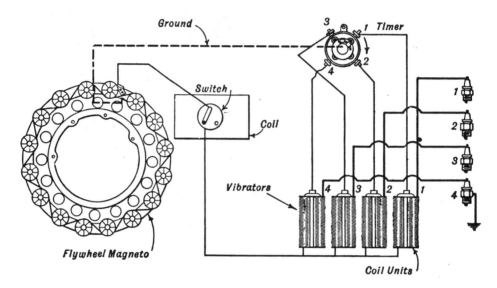

By turning a hand crank, the driver rotated a magneto that sent an electrical current to four coil units (small transformers), which stepped up the voltage. The high voltage went to a distributor (called a commutator in the Ford engine), which sent voltage to each spark plug in rapid sequence, causing ignition of the fuel mixture in the cylinder. Once the engine began running, the crankshaft took over from the hand crank.

Fuel System

Gasoline, stored in the fuel tank, went through a carburetor that mixed the fuel with air. On the intake stroke of the piston, the fuel mixture entered the engine cylinder. Combustion transformed the fuel, represented below by octane (C_8H_{18}) and oxygen (O_2), into the waste products water (H_2O), carbon monoxide (CO), and carbon dioxide (CO_2):[1]

$$C_8H_{18} + 10.5O_2 \rightarrow 9H_2O + 4CO + 4CO_2$$

| Octane | Oxygen | Water | Carbon Monoxide | Carbon Dioxide |

[1] Other elements present in the fuel (e.g., nitrogen) and waste gases (e.g., nitrous oxides) are omitted.

Source: Ford Manual: For Owners and Operators of Ford Cars (Detroit: Ford Motor Company, 1914), p. 45 (ignition diagram).

Sidebar 5.3 **The Model T: Engine and Power**

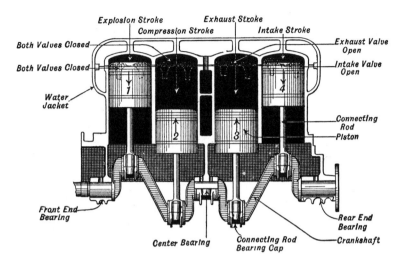

The Ford Model T had a four-cylinder engine that operated on a four-stroke cycle. The diagram shows the position of the pistons at the start of each stroke. The combustion or power stroke is called the "explosion" stroke.

Each cylinder produced one power stroke for every two revolutions of the crankshaft. Thus, in a four-cylinder engine, there were two power strokes per revolution.

The average power on the piston head produced by combustion was the *indicated horsepower* (P_I) of the engine: *PLAN*/33,000. The indicated horsepower of the Ford Model T at its top speed of 37 miles per hour was:

> P = pressure on piston head (70 pounds per square inch, estimated average)
>
> L = length of piston stroke (4 inches or 0.33 feet)
>
> A = area of piston head ($D^2\pi/4$, where D is diameter of the cylinder)
>
> $(3.75^2(3.14)/4 = 11.1$ square inches)
>
> N = number of power strokes (3,000 per minute at 1,500 revolutions per minute)

$$P_I = \frac{PLAN}{33,000} = \frac{(70)(0.33)(11.1)(3,000)}{33,000} = 23.5 \text{ Hp}$$

Sources: Ford Manual: For Owners and Operators of Ford Cars (Detroit: Ford Motor Company, 1914), p. 18 (illustration); *PLAN* data computed from Allan Nevins with Frank Ernest Hill, *Ford: The Times, the Man, the Company* (New York: Charles Scribner's, 1954–63), 1:387–93, and from Floyd Clymer, *Henry's Wonderful Model T, 1908–1927* (New York: McGraw-Hill, 1955), p. 127 (fig. 20).

The traction horsepower had to be sufficient to grip the road and move forward. On an unpaved road, the *rolling resistance* was greater than on a paved surface, but the Model T was built for a nation of mostly unpaved roads and its traction horsepower enabled the car to drive at its top speed of 37 miles per hour. As speed increased, resistance of the air, or *drag*, became greater, but at its top speed the Model T did not encounter significant resistance from the air (sidebar 5.4).[16] The car's gearing system, similar in principle to that of a bicycle, enabled it to climb hills. By changing into a new gear on a bicycle, a rider could shorten or lengthen the distance traveled with each revolution of the pedal. In low gear, each pedal rotation caused the back wheel to turn much less distance than in high gear, which made pedaling easier going uphill. Because the chain transmissions on bicycles were not practical for cars, in its place the rotational power from the crankshaft traveled through a transmission box containing gearwheels of different sizes to a driveshaft running the length of the car.[17] The Model T had two forward gear settings, high and low, and another setting for driving backwards (sidebar 5.5).

With the new car, America had an engineering design that brought together in one vehicle the key innovations of late-nineteenth-century electrical, chemical, and mechanical engineering; the car, in turn, would soon stimulate the civil engineering of new roads and bridges. The simplicity of the numbers that described the car's working parts defined a system of remarkable precision and sophistication. But the car was more than just an efficient machine—it expressed Ford's social vision of a useful vehicle for the great mass of the population. Crucial to this vision was the car's economy; the Model T could not have succeeded as a breakthrough machine without the new process developed by Ford and his engineers to manufacture the car for a mass market.

The Ford Assembly Line

The Model T was an immediate success. But demand soon exceeded Ford's ability to manufacture the car by conventional methods. In common with other car makers in 1908, Ford built his cars from components supplied by others. Each car was assembled in place by workers who moved from car to car on the factory floor. To assemble a chassis required about twelve hours. Ford and his engineers studied the assembly

....

Sidebar 5.4   **The Model T: Traction**

The Model T needed to deliver enough power to the wheels to meet the *traction horsepower* required to overcome the resistance of the road and the resistance of the air at a given speed. The formula for traction horsepower is *TV*/33,000. *T* is for traction force in pounds, and *V* is for velocity in feet per minute.

Traction Force

Resistance of the road, also known as *rolling resistance,* is the product of a road coefficient (C_R) and the weight of the car (*W*). The road coefficient represents the condition of the road surface.

Resistance of the air, or *drag,* is the product of a drag coefficient representing the shape of the car (C_D), the air pressure on the front of the car in pounds per square foot (*p*) at a given speed, and the surface area of the car's front (A_F) in square feet:

C_R = coefficient of road resistance (0.015 for a paved road)
W = weight of car (1,200 pounds for the Model T)
C_D = coefficient of drag (about 1.0 for the Model T)
p = air pressure ($0.00257\,V^2$) (at 37 mph = 3.5 pounds per square foot)
A_F = frontal area of the car (about 28 square feet for the Model T)

Adding rolling resistance and drag gives the traction force (*T*):

$$C_R W + C_D p A_F = T$$
$$(0.015)(1200) + (1.0)(3.5)(28) = 116 \text{ pounds}$$

Traction Power

At its top speed, the Model T had a velocity of 36.86 miles per hour, which we round to 37. To convert *V* into miles per hour, we divide by 88 (88 feet per minute equals one mile per hour) and restate the formula for traction power (P_T) as *TV*/375:

$$P_T = \frac{TV}{375} = \frac{(116)(37)}{375} = 11.4 \text{ Hp}$$

The traction horsepower requirement of 11.4 Hp was less than the 12 Hp that the Model T could deliver to the wheels at 37 miles per hour.

Sources: Transactions of the Society of Automobile Engineers (1915), p. 190 (road coefficient C_R); other data given or computed from Floyd Clymer, *Henry's Wonderful Model T, 1908–1927* (New York: McGraw-Hill, 1955), p. 127 (fig. 20).

....

Sidebar 5.5   **The Model T: Speed**

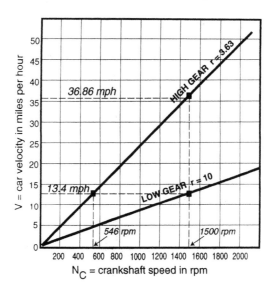

The velocity of the Model T can be calculated from the formula

$$V = (N_c/r)\pi D_w$$

where N_c is the number of crankshaft revolutions per minute (rpm), divided by the gear ratio r, the ratio of crankshaft to driveshaft rpms, and then multiplied by the circumference of the wheel, pi (π), times the wheel diameter, D_w.

The maximum crankshaft speed of the Model T was 1,500 rpm. The gear ratio in high gear was 3.63/1, and the diameter of the wheel was 2.5 feet, giving a velocity:

$$V = (N_c/r)\pi D_w = (1,500/3.63)(3.14)(2.5) = 3,243 \text{ feet per minute}$$

Dividing by 88 to convert feet per minute into miles per hour (mph), the car could travel 36.86 mph in high gear.[1]

In low gear, the car could travel only 13.4 mph at the top crankshaft speed of 1,500 rpm. But at 13.4 mph in high gear, the crankshaft turned only 546 rpm. At 13.4 mph in low gear, the engine delivered more power (1,500 rpm) to the wheels. In low gear at a lower speed, the car could more easily drive uphill or on rough roads.

[1] To convert to mph, 3,243 feet per minute × 60 minutes/5,280 feet per mile = 3,243/88 = 36.86 mph.

Source: Floyd Clymer, *Henry's Wonderful Model T, 1908–1927* (New York: McGraw-Hill, 1955), p. 127 (fig. 20).

process closely. They soon realized that they could accelerate production if the workers stayed in one place and the parts to be assembled came to them by conveyor systems. By the earlier methods, each worker had to perform a multitude of tasks requiring many skills. With a moving assembly, workers could now specialize in a single task, and these tasks could be subdivided. Moving to a much larger plant in the Highland Park section of Detroit in 1910, Ford developed over the next four years a production system consisting of a major assembly line fed by sub-assembly lines on which workers each performed a single repetitive task. The process reduced the time needed to make the chassis from twelve hours to ninety-three minutes (figure 5.7a and b).[18]

The idea of a moving assembly was not new; the meat-packing industry had developed it earlier. What made Ford's assembly line revolutionary was his combination of a moving assembly with another American idea, manufacturing with standardized and interchangeable parts. Ford demanded of his suppliers that every part be machined to high tolerances, so that workers could rapidly assemble cars without having to adjust parts that didn't exactly fit. To ensure high standards, Ford eventually produced all the parts required. The rapid mass production of inexpensive, well-built cars followed. In 1909 the Model T runabout sold for $825. By 1916 the price had fallen to $345. Sales rose from 78,440 cars in 1911–12 to 751,287 in 1916–17 (figure 5.8). Rival auto makers began to imitate Ford's methods, and total car production in the United States rose to 1,745,702 in 1917. By then Ford had captured 43 percent of the market and was by far the largest car company.[19]

To sustain his success, however, Ford had to overcome two challenges: a charge of patent infringement and a high turnover in his labor force. In 1879 the lawyer George Selden of Rochester, New York, had patented an internal-combustion car. By filing a series of amendments, Selden had postponed the commencement of his patent until 1895, by which time it was clear that a market existed for cars. Patent law protected an invention for seventeen years and Selden sued Alexander Winton successfully in 1903 for infringing his claim. The major auto makers then formed an association that paid royalties to an electric vehicle company to which Selden had transferred his patent. Ford refused to join, and the association filed suit. In 1909 a federal district court in New York upheld the Selden patent and threatened to put Ford out of business. An appeals court ruled in Ford's favor in 1911, however, by noting that Selden had only

Figure 5.7a. Ford production: static assembly. Courtesy of The Henry Ford Museum, Dearborn, MI. Photograph no. P.O.1267.

patented a two-stroke engine, not the four-stroke engine used by Ford and most other car makers. With only one more year of the Selden patent remaining, the association declined to take the case to the U.S. Supreme Court and the public hailed Ford as a giant killer.[20]

Labor proved to be a second challenge. The reduction of work to a single task afforded employment to less skilled workers, including large numbers of immigrants from Europe, many black migrants from the American South, and some women and disabled people. But the grinding repetition of the assembly line produced a high turnover each year. Line workers earned about $2.50 a day for a nine-hour day, a rate typical of other firms in the auto industry. In 1914 Ford dramatically raised the wage

Figure 5.7b. Ford production: moving assembly. Courtesy of The Henry Ford Museum, Dearborn, MI. Photograph no. P.O. 3342.

to $5.00 a day for an eight-hour shift. Ford acted from mixed motives. He needed to reduce turnover and he wanted to prevent labor unions from attracting support. But he also wanted workers to benefit from the efficiency of their work and to be able to buy cars. His competitors denounced Ford as a socialist, and again the public hailed him as a man of the people. His acclaim unfortunately proved short-lived.[21]

Alfred P. Sloan and General Motors

World War I (1914–1918), which the United States entered in its last two years, strengthened the demand for motor vehicles but also caused enormous price inflation.

Figure 5.8. One day's production of Model T cars, 1913. Courtesy of The Henry Ford Museum, Dearborn, MI. Photograph no. P.O.716.

By 1920 the dollar had lost half of its purchasing power and the five-dollar day meant less. More seriously for Ford, a great new rival to his company was emerging, General Motors. Under Alfred P. Sloan Jr., who headed the firm in the 1920s, GM displaced Ford as the largest car manufacturer by responding more flexibly to changing demand.

General Motors had its origins in 1904, when a successful carriage maker, William C. Durant (figure 5.9a), switched to cars and took over a small auto manufacturing company founded by David Buick in Flint, Michigan. In 1908 Durant merged Buick with the Oldsmobile firm of Ransom Olds to form General Motors (GM), adding the Cadillac and Oakland (later renamed Pontiac) companies in 1909. By 1910, with its

Figure 5.9a. William C. Durant. Source: *The World's Work*, 40:5 (September 1920): 495.

Figure 5.9b. Alfred P. Sloan Jr. Courtesy of the Hagley Museum and Library, Photographs nos. 69.2, P-SL634-1, PO 96-232.

popular Buick Model 10, General Motors sold more cars than Ford. Durant had financed his expansion with money borrowed from New York banks, however, and a short recession in 1910 caused his bankers to call in their loans. He lost control of GM, which soon fell behind Ford in its share of the market. Durant then launched a new company with the Swiss racecar driver Louis Chevrolet and bought back control of General Motors in 1916, adding the Chevrolet company to its roster. But Durant lost control again in the recession of 1920. At that time, the DuPont family of Delaware had a substantial investment in GM, and Pierre S. DuPont took over as the company's president from 1920 to 1923.[22]

Alfred P. Sloan Jr. (1875–1966) had headed a roller bearing company that General Motors acquired in 1916 (figure 5.9b). In 1918 Sloan joined the managing committee of General Motors, and in 1920 he became the principal assistant to Pierre DuPont.

At Sloan's urging, the auto company made two crucial changes. First, instead of the centralized and rigid management that Ford imposed on his own company, GM created a small executive staff that issued broad numerical targets for sales, market share, and profits, leaving the operating divisions of the company (Cadillac, Buick, Oldsmobile, Pontiac, Chevrolet) wide latitude in meeting them. Second, GM provided consumers with more choice. Each of the five GM divisions served a particular market, with Cadillac providing the most expensive cars and Chevrolet the least costly ones. Most of the working parts of GM cars were standardized but body shape and other visible details varied. Annual style changes also became a feature of GM by the 1930s. Sloan's innovations appealed to the social differences in society that Ford's Model T stood against. But Sloan, who became president of General Motors in 1923, also responded to a public demand for greater variety in cars that Ford was unwilling to meet.[23]

By the 1920s the public had begun to tire of the Model T. Better roads made its ruggedness less attractive. The basic design of the car didn't change, and as sales fell, the company finally phased it out in 1927 and began marketing new cars across a range of prices. By then General Motors had taken Ford's place as the largest U.S. auto company.[24] Ford believed that producing an affordable and reliable car was ethical as well as profitable.[25] But his success depended on rigid standardization, and an autocratic tendency in his personality grew stronger in his later years. In a 1915 libel suit, he declared, "History is more or less bunk." He denounced banks, unions, and jazz music for undermining the small-town way of life, even as his automobiles were helping to end the isolation of rural America. In the early 1920s Ford paid for a series of anti-Semitic articles in a local newspaper that he owned, for which he apologized in 1927. In his later years, Ford exemplified the danger of believing that success in one area of endeavor entitled him to trust his judgment in others. He willed his fortune to establish the Ford Foundation, however, which he left free to support more inclusive goals for society and the world.[26]

Sloan's management of General Motors was not without controversy either. In the early 1920s, the company overruled his advice and produced an innovative air-cooled Chevrolet engine to compete with the water-cooled engines used in other cars. The new engine was rushed into production before its design had been fully worked out.

and tested, and cars with the new engines could not be sold.[27] But Sloan was skeptical as much because the engine was innovative as because the innovation was poorly executed. He believed that the market for cars was maturing and that General Motors would succeed primarily by meeting the need to replace automobiles already in use.[28] Sloan demanded that new cars incorporate improvements only after careful study and agreement within the company. One other innovation by GM in the 1920s, the development of leaded gasoline, led to serious trouble. Cases of lead poisoning among chemical workers who produced the additive raised public concerns that GM and the oil companies tried to dismiss. Public health authorities in the 1920s, however, approved leaded gasoline, which remained in use until environmental concerns led to its phasing out in the 1970s.[29]

Science made little contribution to the automobile in its early years. But the automobile assembly line became an example to many of a new ideal of "scientific management" articulated by Frederick Winslow Taylor. In an influential 1911 book, Taylor argued that industry needed to make every task more efficient by knowing the action of every worker at every moment. To design their assembly line, Ford managers employed time and motion studies of the kind urged by Taylor, and the regimentation of the assembly line caused some social critics to see in modern engineering a totalitarian imperative that robs humanity of its freedom.[30]

There is no evidence that Ford was inspired by Taylor, whose goal was to measure and reward the productivity of individuals. Ford's aim instead was to standardize the output of each worker.[31] Taylor saw only the efficiencies to be squeezed out of a technology that was already established. If Ford had tried to do the same, he would have improved the static assembly of cars instead of abandoning it for a moving assembly line.[32] Taylor's ideal helped encourage a broader misconception that engineering is a matter of using science and mathematics to determine a "one best way." Although it involves calculations, engineering is not just mathematical problem-solving in which there are only right answers and wrong answers. Numbers always constrain what engineering can do but the engineer always has a choice in what to design. The consequences of technological choice are never a matter of inherent or technical necessity, and the idea that an optimal solution exists to every technical problem is a fundamental misunderstanding of modern engineering.

scientific management

101

Americans changed their way of life as they came to rely more and more on automobiles.[33] The car and the road brought together all four principles of network, process, machine, and structure: electricity for ignition, gasoline for fuel, the internal-combustion engine, and the new roads and bridges stimulated and required by the car. After the breakthrough innovation of the Model T, the automotive industry followed the pattern of other industries in shifting, during the 1920s, to the more incremental kinds of change associated with an established technology and market. In the meantime, a second great use for the internal-combustion engine, in the airplane, made an even more radical change in transportation possible: human flight in heavier-than-air craft.

CHAPTER SIX

The Wright Brothers and the Airplane

Successful powered flight began when the Wright brothers flew an airplane on the sand dunes of Kitty Hawk, North Carolina, on December 17, 1903. By the 1930s passenger air travel was still a luxury, but advances in aircraft speed, size, and power were preparing the way for more people to fly, and today millions do. The Wright Flyer, the first powered heavier-than-air craft to fly a pilot in sustained level flight, was the crucial breakthrough that made later aviation possible. Earlier attempts to fly had treated the problem as one of simply getting off the ground.[1] Would-be aviators before the Wright brothers had built wings with engines and had tried to power themselves into the air. Taking a different approach, the Wrights instead studied first how to stay in the air, using gliders. Their focus on how to stay aloft led them to see that an airplane needed maneuverability as well as power. After testing several aircraft designs as gliders, they designed and built a powered airplane and successfully flew it.

Early Attempts to Fly

Modern aviation research began in 1799, when an English landowner, Sir George Cayley (1773–1857), sketched the basic design that airplanes have followed ever since: a fixed lateral wing across a longitudinal body, with a cross-wing tail and a vertical tail rudder. Cayley realized that an airplane would fly if its wings generated *lift*, an upward force counteracting the downward force exerted by the *weight* of the plane. Lift results when the air pressure above the wings is lower than below them. Cayley found that a wing surface generated lift by moving forward with its wing at a positive *angle of attack*, in which the front edge of the wing is elevated higher than the rear edge. At a positive angle, there is more air pressure on the wing's underside than on its upper

103

Figure 6.1. Otto Lilienthal gliding. Source: *National Geographic*, 19:8 (August 1908): 596.

surface. To create enough lift to overcome its weight, an airplane also needs to move forward with a enough force, or *thrust*, to overcome the resistance of the air, or *drag* (sidebar 6.1). Nineteenth-century experimenters later tried to fly powered airplanes with steam engines and propellers, using his basic configuration, but the engines could not produce enough power to overcome their weight at sizes large enough to carry a pilot. Theoretical researchers in aerodynamics took no practical interest in flying.[3]

A German engineer, Otto Lilienthal (1848–96), approached the problem of flight in a new way. Lilienthal studied first what it was like to fly with hang gliders. He believed that successful flight depended on learning how to stay in the air, not just get off the ground. Lilienthal learned from hang gliding that to stay in flight required adjusting the wings to changing air currents, which he did by shifting his weight (figure 6.1). Unfortunately, this method of control limited the size of the glider and was difficult to

....

Sidebar 6.1   **Balancing Forces in Steady Level Flight**

Four Forces and Three Axes

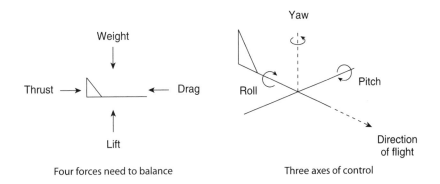

Four forces need to balance   Three axes of control

An airplane maintains steady level flight when its wings generate *lift*, an upward force that overcomes the *weight* of the plane. Lift results when the air pressure below the wings is higher than above them. An airplane also needs to develop enough forward force, or *thrust*, to overcome the resistance of the air, or *drag*.

An airplane also needs to be controlled in three directions. Along the *pitch* axis, the plane can move up or down. Along the *roll* axis, one wing of the plane can dip, causing the other wing to rise. Along the *yaw* axis, the plane can turn left or right.

Angle of Attack

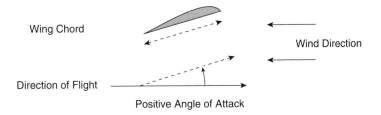

The *angle of attack* is the angle of the wing *chord* in relation to wind direction. The chord is a line that can be drawn between the frontal and rear edges of the wing. At a positive angle of attack, an airplane generates lift by creating more pressure under its wings than above them. The direction of lift is relative to wind direction.

....

Figure 6.2. Samuel P. Langley.
Source: *Cassier's Magazine*, 25:1
(November 1903): 2.

perform in turbulent winds. In 1896 Lilienthal died when one of his hang gliders stalled in the air and fell to the ground.[4]

The leading aviation researcher after Lilienthal was Samuel P. Langley (1834–1906), an American astrophysicist who became secretary of the Smithsonian Institution in Washington in 1887 (figure 6.2). Although not unconcerned about how to stay aloft, Langley focused his attention on how to get off the ground. He conducted careful measurements of the forces acting on various wing shapes and tested small unpiloted models that he called "aerodromes" to experiment with different body and wing configurations. On May 6, 1896, with Alexander Graham Bell present, he launched a model airplane powered by a small steam engine, using a catapult on a houseboat in the Potomac

Figure 6.3. Langley model aerodrome flight in 1896. Source: Samuel Pierpont Langley, *Langley Memoir on Mechanical Flight*, (Washington, DC: Smithsonian Institution, 1911), plate 20 (following p. 108).

River. The unpiloted plane, with a wingspan of thirteen feet, continued to fly on its own power for ninety seconds a distance of half a mile, unprecedented for a model aircraft (figure 6.3). Another model aerodrome flew almost a mile on November 28, 1896.[5]

Langley believed that the next step was to build a larger version of his plane and fly it with a pilot. Believing that it was enough to have proved the concept, he left the practical implementation to others. In 1898, however, the Spanish-American War broke out and the U.S. War Department gave Langley $50,000 to develop a piloted airplane. He continued his research after the war ended later that year. In place of steam power, he substituted a lighter and more powerful gasoline engine built by his assistant, Charles Manly. Langley launched a larger model plane in 1901. A full-scale piloted airplane would be next.[6] Any Americans determined to be the first to fly would have to beat Langley.

Enter the Wright Brothers

Wilbur (1867–1912) and Orville (1871–1948) Wright of Dayton, Ohio, attended high school but did not formally graduate (figure 6.4). They soon educated themselves to a higher level on their own with the help of a home library collected by their father, a bishop in the United Brethren Church, and their mother, who had attended college. The Wrights also learned how to work with their hands from their mother. After failing to sustain a printing business, the Wrights opened a bicycle shop in Dayton in 1892 that was a success. They began making their own line of bicycles a few years later. The Wrights' combination of manual skill and intellectual curiosity eventually drew them to the challenge of powered flight.[7]

On May 30, 1899, Wilbur Wright wrote to the Smithsonian Institution for information on aviation. An assistant to Langley replied with several pamphlets and a list of suggested readings. These included *Progress in Flying Machines*, an 1894 book by Octave Chanute, a civil engineer who had become a clearinghouse of information on aviation in the United States. Chanute would later give the Wrights advice and encouragement.[8] With the help of Chanute's book and the other material, Wilbur and Orville quickly brought themselves up to date on the work of earlier researchers. They decided that, despite his accident, Otto Lilienthal had been correct to try to learn how to stay in the air before attempting powered flight to get off the ground. But the Wrights realized that the key to staying in the air was to have better control over the aircraft, so that it could adjust to changes in the wind.[9]

To stay in level flight, a pilot needs to control the rotational motion of an airplane around three axes (sidebar 6.1). The first, defined by a line parallel to the wingspan, is called the *pitch* axis. If the airplane rotates around it, the plane will head up or down. The second axis runs lengthwise through the fuselage and is known as the *roll* axis. If a side wind pushes one wing up, causing the other to go down, the airplane will roll or rotate around this line. If a plane in level flight turns to the right or left, like a car turning on a road, it is said to *yaw* or turn on its vertical axis.[10] Aviation researchers before the Wright brothers had recognized the need for stability in flight, but most saw it as a problem to be solved in the design of the airframe. Using a model airplane with a rubber-band propeller, Alphonse Pénaud of France had found that a tail wing set at a

Figure 6.4. Wilbur and Orville Wright in 1909. Courtesy of the Still Picture Branch, National Archives, College Park, MD. RG237-G-196-21.

slight negative angle to the wind would help stabilize the plane with its main wing at a slight positive angle. He also made the wing tips slightly higher than the places where they attached to the main body or fuselage, giving some stability in roll. But Penaud's wings were fixed in position and thus could not be controlled flexibly.[11]

The Wrights saw the problem of stability very differently. While an airplane required a design that had some inherent stability, it also needed flexible control so that a pilot could overcome the effects of unexpected turbulence. Because the Wrights did not see weight shifting as a practical way to control an airplane in flight, they designed another form of control that they tested with a five-foot-long biplane kite in the summer of 1899. The kite had hand-held wire controls that could pull or "warp" the rear edge of each wing tip up or down. By raising or lowering the right and left rear edges together, the Wrights could make the kite fly up or down in the direction of its pitch. By raising the rear edge of the right wings and lowering the left ones, or vice versa, they could also bank the kite to the right or left, just as a cyclist did when turning.[12] The kite tests encouraged the Wrights to build a glider and gain experience flying with these wing controls.

The Wright Gliders

The Wright brothers had to support their research out of their income as makers and sellers of bicycles, and their business usually occupied them except in the autumn. But the brothers found the time to complete a design and make parts for their first glider by August 1900. The Wrights traveled in September to the village of Kitty Hawk, North Carolina, where steady onshore breezes from the ocean and enormous empty beaches and sand dunes created a relatively safe place to conduct glider flights.[13]

The first Wright glider consisted of biplane wings, seventeen and one-half feet long and five feet wide, made of curved wooden slats and sateen fabric. Vertical struts held the two wings together and diagonal wire trusses provided reinforcement in the front and back but not on the sides, so that the rear edges of the wings could be pulled up or down using wire controls. Sir George Cayley had found that giving the upper surface of a wing a slight upward curvature or *camber* caused air to move faster over the wing than under it. The result was less air pressure above the wing and thus greater lift. The Wright broth-

Figure 6.5. The Wright 1900 glider flying as a kite. Courtesy of the Prints and Photographs Division, Library of Congress, Washington, DC. LC-W851-115.

ers gave their wings a slight camber. The brothers also placed the tail wings in front of the main biplane wings, rather than behind them, to lessen the danger of stalling.[14]

With the help of local villagers, the Wrights began to test their glider, at first tethered to the ground (figure 6.5) and then in free flight. In the first free flights, Orville and Bill Tate, a village youth, held each end of the wings with Wilbur lying prone on the lower wing as pilot. The breeze lifted the plane and carried it several hundred feet. The Wrights tested the drag of the airframe by tethering it to a weighing scale and measuring the pounds of force exerted by the wind. Drag turned out to be low. But lift also turned out to be low. The Wrights had calculated that their plane would lift itself and a pilot in a twenty-one-mile-per-hour wind at a positive angle to the wind of three degrees. At Kitty Hawk, measurements indicated that the glider would lift this weight

Figure 6.6. The Wright 1901 glider upended. Courtesy of the Prints and Photographs Division, Library of Congress, Washington, DC. LC-W851-96.

only at an angle of twenty degrees in a twenty-five-mile-per-hour wind. An angle of attack this high would have unacceptable drag.[15]

Over the following winter and spring, the Wright brothers built a new glider with the main wings measuring twenty-two feet by seven feet and with a higher wing camber (figure 6.6). Returning to Kitty Hawk in July 1901, the Wrights began making free flights again (figure 6.7). But this time, their glider became unstable in pitch and began stalling. The airframe landed safely but the glider was less stable than the previous year's plane. When the brothers tethered the glider and flew it as a kite to measure lift, they also found that the larger airframe generated less lift than their previous glider. By adjusting the wings on the spot to reduce the camber, the Wrights were able to restore pitch stability to the plane. Then a more serious problem emerged. In several tests of

Figure 6.7. Wilbur Wright flying the 1901 glider. Courtesy of the Prints and Photographs Division, Library of Congress, Washington, DC. LC-W851-123.

the wing controls, Wilbur banked the plane in one direction only to have the wings suddenly move in the opposite direction. The Wrights' great insight of flexible wing controls now appeared to be wrong. The Wright brothers returned home deeply discouraged.[16]

Rethinking the Fundamentals of Flight

The Wright brothers were not down for long. In the spring of 1900, they had begun to correspond with Octave Chanute, who encouraged them in their efforts. After the

Wright brothers returned home in late August 1901, Chanute invited Wilbur to report their research to the Western Society of Engineers in Chicago on September 18.[17] Wilbur's slide lecture was concise and well received, and the interest of professional engineers restored the confidence of the brothers in their quest to fly. But the Wrights now realized that something was fundamentally wrong.

In designing their plane, Wilbur and Orville had relied on earlier research to calculate lift at various speeds. The basis of this research was the work of the eighteenth-century English engineer John Smeaton on flows of water and wind against flat surfaces. Smeaton calculated the pressure of a flow perpendicular to a flat plate with the formula $F = kV^2S$, in which F is the force hitting the plate (in pounds), kV^2 is the air pressure in pounds per square foot, and S is the surface area of the plate, measured in square feet. In the number for air pressure, V is the velocity of the air in miles per hour, and k is a number, known as Smeaton's coefficient, that in part represents air density. Smeaton assigned this coefficient a value of 0.005 (sidebar 6.2).[18]

To calculate lift and drag on a wing at a given angle of attack, aviation researchers added a coefficient for lift or drag to Smeaton's formula. Adding a coefficient for lift (C_L) defined lift as: $L = kV^2SC_L$, with L the lifting force and S the surface area of the wingspan. Substituting a drag coefficient (C_D) for the lift coefficient defined drag as: $D = kV^2SC_D$, with D the drag force. The value of each coefficient for any given wing shape had to be determined for each angle by testing. Otto Lilienthal had studied lift and drag on different wings and had produced tables of coefficients at various angles of attack. These were the best tables in existence and the Wright brothers relied on them and on Smeaton's coefficient to design their gliders.

The Wrights suspected that Lilienthal's numbers for lift and drag were mistaken. But finding the correct numbers was not easy, because the formulas for each contained two coefficients that were now uncertain: one for lift or drag and the other Smeaton's coefficient. The Wrights built a small wind tunnel in October 1901, out of a wooden crate, in which they placed model wing surfaces. The brothers skillfully found ways to test lift and drag numbers independently of Smeaton's coefficient. They initially concluded that Lilienthal's numbers were inaccurate until they realized that the numbers

The Problem of Lift

The Wright brothers learned how to design an airplane by testing full-scale gliders. Their 1900 glider did not produce enough lift, so in their 1901 glider they lengthened and widened the wings and increased the wing camber. To their surprise, the glider performed far less well. Suspecting an error in the numbers they had used to calculate lift and drag, they built a wind tunnel to test wing shapes.

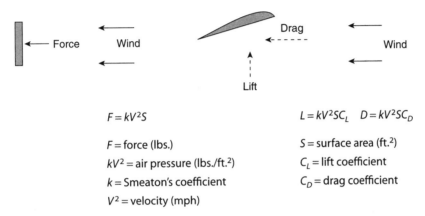

$$F = kV^2S$$

$$L = kV^2SC_L \quad D = kV^2SC_D$$

F = force (lbs.)

kV^2 = air pressure (lbs./ft.^{2})

k = Smeaton's coefficient

V^2 = velocity (mph)

S = surface area (ft.^{2})

C_L = lift coefficient

C_D = drag coefficient

The flow of air against a flat plate can be expressed as a force $F = kV^2S$. By adding a coefficient for lift or drag to the formula for force, the lift and drag can be calculated for different angles of attack.

The value of each lift and drag coefficient has to be determined by testing for each wing shape at specific angles of attack. The Wright brothers tested wing shapes in a homemade wind tunnel. Their results for Wing Surface No. 12 are given below:

Angle of Attack (in degrees)	C_L (lift)	C_D/C_L	C_D (drag)	Velocity (mph)
0	0.145	0.263	0.038	51.3
2.5	0.311	0.138	0.043	35.0
5.0	0.515	0.105	0.054	27.2
7.5	0.706	0.108	0.076	23.2
10.0	0.839	0.118	0.099	21.3

Source of C_L and C_D/C_L values: Marvin W. McFarland, ed., *The Papers of Wilbur and Orville Wright* (New York: McGraw-Hill, 1953), 1:579, 583 (for Surface No. 12). C_D computed from C_L and C_D/C_L ratios. Velocity computed from $L = kV^2SC_L$, with $L = 625$ lbs., $k = 0.0033$, and $S = 500$

may have been correct for the wing shape Lilienthal used, which was different from the shape the Wrights had used in their own gliders. The Wrights tested many wing shapes and compiled a new set of lift and drag coefficients over a range of angles to the wind.[19]

The Wright brothers soon realized the source of their difficulty: Smeaton's coefficient was wrong. The Wrights concluded from their wind tunnel research that the factor k should be 0.0033 rather than Smeaton's number 0.005.[20] Modern research has shown that the value of k at sea level is 0.00257, but the new factor the Wrights used would be accurate enough for their purposes. The wind tunnel results also helped the Wright brothers find more efficient dimensions for the wings. They determined that the wings needed a lower camber, not a higher one. Studying the wing aspect ratio, or ratio of length (from tip to tip) to the width, the brothers also found that a longer and narrower wing would generate more lift than a shorter and wider wing having the same surface area. The Wrights solved the problem of banking the plane in flight: in banking one wing to turn, the other wing acquired more drag. By using a vertical rudder behind the airplane, they believed that this effect could be counteracted.[21]

The Wrights built a new glider with main wings thirty-two feet long and five feet wide, and a vertical rudder extending out from the rear. In September 1902 the brothers returned to Kitty Hawk. In free flights, the new plane proved itself dramatically (figure 6.8). The glider flew at a seven degree angle of attack at the desired minimum windspeed of twenty-three miles per hour. The Wrights needed to make the vertical tail rudder adjustable, but with that modification the brothers believed they were now ready to make a powered airplane.[22]

The Wright Flyer

In their design for a powered plane, the Wrights aimed for a total weight of airplane, engine, and pilot of 625 pounds. For the shape of the main biplane wings, they chose one of the airfoils (wing shapes) that they had tested in their wind tunnel, and they planned the wings to measure 40 feet by 6.25 feet, or 500 square feet in total area. The

Figure 6.8. Wilbur Wright banking the 1902 glider. Courtesy of the Prints and Photographs Division, Library of Congress, Washington, DC. LC-W861-7.

airplane would need to stay within a range of attack angles that they considered safe, between 2.5 and 7.5 degrees: lift would be difficult to achieve at a lower angle of attack and a higher angle would risk stalling. The need for speed would be greatest at the lowest angle, where there was the least lift. At 2.5 degrees, the required speed would be 35 miles per hour (sidebar 6.2).[23]

Thrust is measured in pounds, and to achieve a velocity of 35 miles per hour, the Wrights calculated that they would need an airplane with a thrust of 90 pounds (sidebar 6.3). The number for thrust enabled the Wright brothers to calculate the horsepower they would require in their engine. Multiplying thrust in pounds by velocity in miles per hour, and dividing by 375, gave the horsepower. With 90 pounds of thrust and a speed of 35 miles per hour, the engine would need a horsepower of 8.4. The Wrights could not find a shop able to make an internal-combustion engine with the required power at a low enough weight. So with the help of Charles Taylor, their bicycle shop assistant, the brothers built a four-cylinder gasoline engine themselves that weighed under 180 pounds. The brake horsepower of the engine came to 11.81.[24]

The ability to fly depended, however, on how much thrust the airplane would have after any power losses occurred in the transmission and the propellers. In trying to calculate performance, the Wright brothers soon realized how difficult it would be to calculate the thrust of a propeller on a moving airplane compared to a stationary wing in a wind tunnel:

> What at first seemed a simple problem became more complex the more we studied it. With the machine moving forward, the air flying backward, the propellers turning sidewise, and nothing standing still, it seemed impossible to find a starting-point from which to trace the various simultaneous reactions. Contemplation of it was confusing. After long arguments, we often found ourselves in the ludicrous position of each having been converted to the other's side, with no more agreement than when the discussion began.[25]

Eventually, the Wrights decided to treat the propeller as they would a wing and to design it as if it was producing lift. The brothers estimated a power loss in the propellers of 66 percent and a loss of 5 percent in the chain transmission from the engine, which would not have given them enough power to fly 35 miles per hour at a 2.5 degree angle of attack. The Wrights would need to fly at a higher angle.[26]

The Wright brothers brought their plane to North Carolina in September 1903. The higher brake horsepower of the engine had encouraged the brothers to add weight in the form of strengthening to the airframe, bringing the total weight (with pilot) up to 750 pounds. The main wingspan had also grown from 500 to 510 square feet. Octave Chanute visited them at Kitty Hawk and warned that power losses from the engine to the propellers might be greater than the Wrights had planned. The airplane was now designed to include almost no margin for error. Anxiously, the brothers tested the plane with a rope and pulley to determine how many pounds of sand it

. . . .

Sidebar 6.3   **The 1903 Wright Flyer: Design**

Calculating Lift and Drag

The Wrights determined that they would need to fly between attack angles of 2.5 and 7.5 degrees. They designed their plane to fly at a worst-case angle of 2.5 degrees, at which the plane would need the greatest speed, 35 miles per hour (see sidebar 6.2).

The Wright brothers expected their plane to lift a total weight (airplane and pilot) of 625 pounds. From the lift coefficient for their chosen wing shape (No. 12) at 2.5 degrees, they believed that a wing area of 500 square feet would be sufficient, assuming a value for the coefficient k of 0.0033. The designed lift (L) came to slightly above 625 pounds:

$$L = \quad kV^2 \qquad S \qquad C_L$$
$$628 = 0.0033(35)^2 \quad (500) \quad (0.311)$$

The Wrights then calculated how much drag (D) there would be on the airplane, using the drag coefficient for wing No. 12 at 2.5 degrees (sidebar 6.2). To calculate drag, they added the frontal area of the plane (S_F) to the surface area of the wingspan (S), an addition of 20 square feet. The drag gave the thrust (T) needed at 35 miles per hour, 90 pounds:

$$D = \quad kV^2 \qquad (S + S_F) \qquad C_D = T$$
$$T = (0.0033)(35)^2 \; (500 + 20) \quad (0.043) = 90 \text{ lbs.}$$

From Thrust to Power

The Wrights then calculated that they would require an engine with 8.4 horsepower to achieve 90 pounds of thrust. The formula for traction power in an automobile could be used to calculate the thrust power in an airplane, where P_T is the power, T stands for thrust instead of traction, and V is velocity in miles per hour:

$$P_T = \frac{TV}{375} = \frac{(90)(35)}{375} = 8.4 \text{ Hp}$$

·

Source: Howard S. Wolko, "Structural Design of the 1903 Wright Flyer," in Howard S. Wolko, ed., *The Wright Flyer: An Engineering Perspective* (Washington, DC: Smithsonian Institution Press, 1987), pp. 98–100.

. . . .

·

could pull. It soon became clear that the propellers could deliver 132 to 136 pounds of thrust, more than enough to overcome any power losses in the transmission and propellers (and the error, unknown to them, in their estimate of the air density factor, k).[27]

While the Wrights prepared for their test flight, Samuel Langley in Washington prepared to fly his own full-scale airplane. With the assistance of Charles Manly and a staff of workers, Langley had designed an airplane without first acquiring any experience gliding in the air. The gasoline engine designed by Manly weighed 124 pounds and delivered a huge 52 horsepower. The "Great Aerodrome" had two single wingspans, one in front of the propellers and one behind them, with a smaller rear tail wing and rudder at the end. Manual controls for moving the tail wing and rudder controlled pitch and yaw. There was no manual control over movement in roll. Langley tried to design the aircraft to be passively stable in flight. Reducing as much as possible the need for human control, he thought, would make the aircraft easier and safer to fly.[28]

Langley tests his plane

On October 7, 1903, Langley's plane was ready. It took off by catapult over the Potomac River and immediately plunged into the water with Manly on board. But Manly and Langley blamed the launch mechanism for the failure and began preparing for a second flight. On December 8, 1903, the Great Aerodrome was reset on its houseboat catapult late in the day under blustery conditions. The catapult propelled the aircraft along a sixty-foot track. Upon leaving the platform, the airplane pitched up into the air at a 90 degree angle and collapsed into the river. Langley and his friends in the scientific establishment blamed the launch mechanism again. But Langley received no more public funding to conduct research, and he died in 1906.[29]

The Wright brothers at Kitty Hawk endured troubles of their own: in early November a propeller shaft broke, and no sooner had a replacement arrived by the end of the month than another shaft broke. Orville returned to Ohio to make new and stronger shafts and returned on December 11. The brothers were finally ready. Wilbur won a coin toss and piloted the first flight on December 14. Upon leaving the ground, the airplane pitched up too high, stalled, and came down after a flight of 60 feet. The Wrights decided that this was not a true flight, so after making some repairs Orville took the plane on December 17 (sidebar 6.4 and figure 6.9). He described what followed:

After running the motor a few minutes to heat it up, I released the wire that held the machine to the track, and the machine started forward into the wind. Wilbur ran at the side of the machine, holding the wing to balance it on the track. Unlike the start on the 14th, made in a calm, the machine, facing a 27-mile wind, started very slowly. Wilbur was able to stay with it till it lifted from the track after a forty-foot run.

One of the Life Saving men snapped the camera for us, taking a picture just as the machine had reached the end of the track and had risen to a height of about two feet. The slow forward speed of the machine over the ground is clearly shown in the picture by Wilbur's attitude. He stayed along beside the machine without any effort. The course of the flight up and down was exceedingly erratic, partly due to the irregularity of the air, and partly to lack of experience in handling this machine.

The control of the front rudder was difficult on account of its being balanced too near the center. This gave it a tendency to turn itself when started; so that it turned too far on one side and then too far on the other. As a result the machine would rise suddenly to about ten feet, and then as suddenly dart for the ground. A sudden dart when a little over a hundred feet from the end of the track, or a little over 120 feet from the point at which it rose into the air, ended the flight.

As the velocity of the wind was over 35 feet per second and the speed of the machine over the ground against this wind ten feet per second, the speed of the machine relative to the air was over 45 feet per second, and the length of the flight was equivalent to a flight of 540 feet made in calm air. This flight lasted only 12 seconds, but it was nevertheless the first in the history of the world in which a machine carrying a man had raised itself by its own power into the air in full flight, had sailed forward without reduction of speed, and had finally landed at a point as high as that from which it started.[30]

Sidebar 6.4 **The 1903 Wright Flyer: Performance**

The 1903 Wright Flyer

The Wright brothers built an engine that weighed about 180 pounds and achieved a brake horsepower of 11.81. The horsepower encouraged the Wrights to add strength to their airframe, bringing its weight (with pilot) up to about 750 pounds. The wingspan also grew from 500 to 510 square feet, with another 20 square feet in frontal area.

A transmission loss (0.95) and a further efficiency loss in the propellers (0.66) reduced the horsepower from 11.81 to 7.4 at the propellers. But the thrust of the engine, when tested at Kitty Hawk, turned out to be 132 to 136 pounds, more than enough to fly the plane.

The Wright Brothers' First Flight

The headwinds on December 17, 1903, averaged about 24 miles per hour. At 10:35 AM, Orville Wright piloted the plane 10 feet off the ground and flew 12 seconds over a distance of 120 feet, at an airspeed of 45 feet per second or 30.7 miles per hour.

To achieve lift at this speed, the coefficient of lift (C_L) had to have been about 0.610 (sidebar 6.2), suggesting an angle of attack of about six degrees. At this angle, the coefficient of drag (C_D) was probably about 0.065. With k at its true value of 0.00257, the airplane needed a drag or thrust (T) of 84 pounds:

$$T = \quad kV^2 \qquad (S + S_F) \qquad C_D$$
$$84 \text{ lbs.} = (0.00257)(30.7)^2 \quad (510 + 20) \quad (0.065)$$

The thrust power (P_T) required to fly that morning, $TV/375$, was 6.9 horsepower. Even if the propeller horsepower had been only 7.4, there was sufficient power for the flight:

$$P_T = \frac{TV}{375} = \frac{(84)(30.7)}{375} = 6.9 \text{ Hp}$$

Source: Omega G. East, *Wright Brothers* (Washington, DC: National Park Service, 1961; repr., 1991), pp. 36–39. Flight data from Marvin W. McFarland, ed., *The Papers of Wilbur and Orville Wright* (New York: McGraw-Hill, 1953), 1:395n.

Figure 6.9. Orville Wright making the first powered flight, December 17, 1903. Courtesy of the Prints and Photographs Division, Library of Congress, Washington, DC. LC-W861-35.

Three more flights took place that day, ending with one by Wilbur that covered 852 feet in 59 seconds. Later that afternoon, a gust of wind overturned the parked airplane and broke the airframe. But the Wright brothers had achieved their goal.[31]

The Wright Brothers: Aftermath

The Wrights attracted some publicity at the time of their first flight, but a new plane they built the following year proved harder to fly. The brothers perfected their design in 1905, though, enabling Wilbur to stay up for 39 minutes and cover a distance of 24.5 miles around Dayton. The Wrights filed a patent on their airframe and control system in the United States that was granted in 1906.[32]

The Wright brothers tried to sell their plane to the U.S. Army. To their surprise, they were turned down. The Army had backed Langley's airplane research and had come under criticism when the Great Aerodrome had failed. Military leaders wanted to examine detailed drawings and see an airplane fly before signing a contract. The Wrights did not believe that they could defend their patent in court against better-financed competitors, though, and they feared that a public demonstration would lead to the pirating of their design. The British, French, and German governments also refused to sign contracts until they had seen a demonstration of the Wright Flyer. For several years, from 1905 to 1908, the Wright brothers kept their invention to themselves. Aviators in France soon began to fly inferior planes based on reports of what the Wright Flyer looked like, although the French planes could not bank or travel distances longer than a few hundred feet.[33]

A more serious competitor appeared in North America, sponsored by Alexander Graham Bell, who in later life conducted aviation experiments on his estate in Nova Scotia, Canada. For reasons of safety, Bell believed that the future of aviation lay in slower airplanes, not faster ones. He invented a multiplane wing that was stable in a 10 mile per hour wind and he designed an airplane to travel at that speed.[34] Bell's plane never flew, but he attracted a staff of younger assistants, including a motorcycle racer and engine maker, Glenn Curtiss, who joined Bell in 1907. With Curtiss, Bell's staff developed a faster plane. On July 4, 1908, Curtiss flew the plane more than 5,000 feet in a contest sponsored by the magazine *Scientific American*.[35]

The Wright brothers had to act. In the summer of 1908, Wilbur Wright took the 1905 airplane to France and gave a public demonstration in August. His dramatic performance completely outclassed the clumsier machines flown by his European rivals. Repeating his flights over several days to ever-increasing crowds, and flying again in late fall, he stunned the public and became the most sought-after celebrity in Europe. Wilbur quickly formed private airplane manufacturing companies with local investors in France, England, and Germany. In September 1909, he flew over the Hudson River in New York City for the 300th anniversary of Henry Hudson's arrival and (two years late) the 100th anniversary of Robert Fulton's steamboat trial of 1807. Wilbur circled the Statue of Liberty to cheering crowds and saluting ships in the river (figure 6.10). While Wilbur was in Europe, in August and September 1908, Orville Wright went to

Figure 6.10. Wilbur Wright circling the Statue of Liberty, 1909. Source: *Harper's Weekly*, 53:2755 (October 9, 1909): 3.

Fort Myer, Virginia, to demonstrate another 1905 plane to officers of the U.S. Army. Despite a crash landing in which an officer was killed, the trials were successful and the Army signed a contract. The brothers formed an American company with investors in New York to manufacture airplanes, and Orville personally trained the first military pilots.[36]

Over the next few years, however, the Wright brothers failed to design and sell competitive airplanes for a larger market. As bicycle makers, they had produced all of their bicycles by hand, and their demonstration airplanes were also hand-made. Like Bell, their talent had been to create a single breakthrough invention, not to run a large business. The planes made by their company were difficult to fly and began to lose their technological edge. As the Wright brothers had feared, public demonstrations of their plane enabled rivals to copy its most useful feature, its principle of flexible wing controls. Competing aircraft makers replaced wing warping with ailerons, rigid flaps built into the rearward edge of each wing that could be raised or lowered without bending the rest of the wing. New airplanes also mounted the engine and propeller in front. In place of the lever-operated controls on the Wright planes, new aircraft began to use a much simpler single control stick that could be moved forward and sideways.[37] The Wrights had patented their airplane in such broad terms that any airplane with flexible controls infringed it. The Wright brothers, in effect, claimed a world monopoly over flyable planes. But European aviators infringed the patent, claiming loopholes, and the German Wright patent was overturned altogether. The Wright company in France was poorly managed and made no money for the brothers.[38]

Glenn Curtiss left Bell to form an aircraft manufacturing company in 1909 only to be sued and put out of business by the Wright brothers in 1910. But an appeals court allowed Curtiss to reorganize his company and sell seaplanes to the U.S. Navy. Henry Ford saw the Wright patent as the kind of stranglehold that he had fought in the Selden case, even though Selden's invention hardly compared to the Wright Flyer, and Ford gave his attorney to Curtiss. In early 1914 the appeals court ruled in favor of the Wright patent. But Orville (Wilbur had died in 1912) had no interest in exploiting his monopoly. He sold the Wright company in 1915 and devoted himself to laboratory research. The Wright company entered a patent-sharing agreement with its competitors in 1917.[39]

Following the Wright legal victory in 1914, Curtiss found a way to strike back. Charles Walcott, Langley's successor as secretary of the Smithsonian Institution, believed that Langley deserved the credit that had gone to the Wright brothers. He authorized Albert Zahm, director of the Smithsonian's Langley Aeronautical Laboratory, to rebuild the Langley plane and show that it could have flown. Zahm engaged Curtiss to fly it, and the two made a number of crucial and unreported changes that strengthened the original airframe. Curtiss flew the rebuilt plane over Lake Keuka, near his headquarters in Hammondsport, New York, in May 1914. The Smithsonian returned the plane to Washington, DC, and declared it the first true airplane.[40]

Orville Wright was understandably shocked. When photographs of the original and rebuilt Aerodrome proved that the rebuilt plane was clearly a different aircraft, Orville demanded a retraction and apology. Walcott and his successor refused. In 1928 Orville sent the 1903 Wright Flyer, which had been restored, to the Science Museum in Kensington, England, with orders that it remain out of the United States until the Smithsonian admitted its wrong. The exile of the Wright Flyer became an increasing embarassment to the Smithsonian and to the nation. Finally in 1948, after Orville's death, the Smithsonian retracted its claim and negotiated with the Wright estate for the Flyer's return. It now hangs in a place of honor above the entrance to the National Air and Space Museum, where a plaque recognizes it as the world's first successful powered airplane.[41]

The scientific field most relevant to aviation was fluid dynamics, the study of flows, which had made important theoretical advances by the late nineteenth century. Leonhard Euler (1707–1783) had formulated equations for modeling a frictionless flow, and the Navier-Stokes equations worked out in the 1840s had modeled flows involving friction. The 1880s brought an improved understanding of the transition from smooth to turbulent flows. "On the other hand," wrote the historian John D. Anderson, "transfers of technology from that advanced state of the art in fluid dynamics to the investigation of powered flight were virtually nonexistent." Aerodynamic theorists were content to leave the practical problem of flying to people whom they regarded as mere technicians or as madmen, and the equations for flows were useless to any practical effort to design an airplane. Lilienthal and Langley could see no use for them, and, as Anderson wrote, "No theoretical flow-field solutions stemming from the Euler equations or the Navier-Stokes equations were used in the design of the

Wright Flyer."[42] Gustave Eiffel (1832–1923) of France, the famous civil engineer, conducted wind-tunnel research that extended the work of the Wrights, and Ludwig Prandtl (1875–1953) of Germany, along with other theorists, soon found ways to describe airflows over wings which had more practical guidance to engineering designers. Theoretical research useful to aviation came only after the Wright brothers' achievement—not before their breakthrough.

Langley was the first scientist of this generation to take a serious interest in flight. However, the aim of his research was limited to making an airplane inherently stable and thus to minimize the need for human control. This approach was appropriate for flying model planes, but not for piloting a full-scale airplane. Langley spent federal grant money on a sophisticated research program, but an urgent desire to get off the ground and into the air led him to take risks with bad weather, using a launch system that had failed once before. The result was his second and final crash.

The Wright brothers studied earlier attempts to fly and benefited from the work of Cayley and Lilienthal. When the Wrights encountered difficulties, they reexamined their basic assumptions with the help of careful wind tunnel tests—although their calculations were not complex. The Wrights designed and carried out a program of field testing under conditions as realistic as possible, and they learned to fly in gliders before trying to pilot a more complex powered airplane. They owed their success to an early insight: the way to achieve stability was to maximize human control of the aircraft. They were more successful as engineers than as entrepreneurs, and other designers carried new technology of powered aviation forward (see Chapter 10).

CHAPTER SEVEN

Radio: From Hertz to Armstrong

*J*ust as the airplane released transport from the confines of rail and road, the innovation of radio freed communication from wire networks. When it emerged at the turn of the century, radio was used to provide a new form of telegraphy. Radio engineers soon began to use it for broadcasting, making communication possible from one point to millions. By the 1920s radio stations began to serve a popular audience for news, music, sports, and other forms of entertainment. Radio technology did not advance smoothly, though, and pride and profit turned some of the leading engineers and entrepreneurs of radio against each other. But radio defied predictions that it would disappear as the new medium of television emerged, and its use of electronics launched the modern electronics industry.

Electromagnetic Waves

The electric power network used electromagnetism in transformers to step voltage up and down, and the telephone used electromagnetism to vibrate a diaphragm in the receiver. Radio takes advantage of an additional feature of electromagnetism. When a current begins to flow, and when there is a change in its flow, it emits an electromagnetic wave that can be detected over a much greater distance than the magnetic field produced by a steady current. No one knew that such electromagnetic or radio waves existed until the Scottish physicist James Clerk Maxwell (1831–1879) proposed them and calculated that they would travel at the speed of light.[1] In 1887 a young German physicist, Heinrich Hertz (1857–1894), designed and tested an apparatus to try to confirm the existence of these waves (sidebar 7.1).[2]

Hertz built a transmitting antenna that consisted of two wire lengths separated by a three-quarter centimeter gap. He connected the two lengths to a source of pulsed high

Sidebar 7.1 **Hertz and Radio Waves**

Electromagnetic Waves

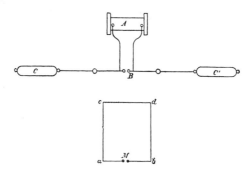

When an electric current flows in a circuit, a magnetic field forms around the current that diminishes with distance. The oscillation of the current emits an electromagnetic wave that travels over a far greater distance.

Heinrich Hertz measured the length of an electromagnetic wave in his laboratory by sending pulsed high voltage from an induction coil *A* to a dipole transmitting antenna *CC'* (above). At high enough voltage, a spark formed in the gap of the transmitting antenna, allowing an alternating current to flow across the antenna. With each oscillation of the current, the antenna emitted a wave. The wave was strong enough to cause the gap in the receiving loop *acdb* to spark.

Hertz placed a metal surface behind the receiving loop that reflected waves back towards the antenna. He then moved the receiving loop towards the reflector and noted the brightness of the spark, which was brightest at intervals of 4.65 meters. Since each wave peaked twice, Hertz measured the wavelength at 9.3 meters.

The frequency of an electromagnetic wave is given by the formula $f = c / \lambda$, where *f* is the frequency in cycles per second (cycles per second are now known as *H* or hertz), *c* is the speed of light in meters per second, and λ (lambda) is the length of the wave in meters. Assuming the speed of light to be 300,000,000 meters per second, Hertz calculated that a wavelength of 9.3 meters would give a frequency of 32,300,000 cycles per second, or 32.3 MHz (megaherz).

To confirm his finding, Hertz also calculated the frequency from measurements of his dipole antenna. This calculation produced a slightly higher frequency of 35.6. But this result was close enough to confirm the existence of electromagnetic waves that traveled at the speed of light.

Source: Heinrich Hertz, *Electric Waves* (1893, repr. 1962), p. 40 (illustration).

voltage. With each pulse, electrons jumped across the gap in a brilliant spark that allowed an alternating current to flow across the antenna. With each oscillation of the current, Hertz believed that the antenna would propagate electromagnetic waves. To detect the wave, Hertz placed a loop of wire of the same length as the antenna a few meters away. The loop also had a small gap in it. He observed that the gap in the loop sparked when the antenna did, proving that electromagnetic waves had traveled through the air.

But to confirm that these were Maxwell's waves, Hertz needed to measure their frequency, length, and speed. The frequency of a wave (f) could be found by dividing its speed by its length. To find the length (λ, lambda), Hertz placed a metal reflector behind the receiving loop, facing the transmitter. The reflector caused a *standing wave*, in which the peaks of the emitted and reflected waves were in step with each other at fixed distances from the transmitter.[3] By moving the receiving loop until the spark was brightest, Hertz found one of these peaks. He found the next by moving the loop farther away from the transmitter until it was brightest again. After several trials, he measured the average of the distance between each peak to be 4.65 meters. Because each wave peaked twice, as it went positive and then negative, Hertz's wave was therefore about 9.3 meters long. Maxwell had proposed that electromagnetic waves traveled at the speed of light, c, which scientists took to be 300 million meters per second. Solving the formula $f = c/\lambda$ with these numbers gave a frequency f of 32.3 million cycles per second. But Hertz needed to confirm independently that the wave traveled at the speed of light. A second formula enabled him to do so.

Direct and alternating current normally encounter resistance in a circuit, and alternating current encounters two more impediments called *reactances*. The ability of a circuit to store electric and magnetic energy (see chapter 3) creates these impediments. The ability to store electric charge, the *capacitance* (C), creates a resistance to change in the voltage that is measured in units called *farads* (named for Michael Faraday). The ability to store magnetic energy, the *inductance* (L), creates a resistance to change in current flow that is measured in units called *henries* (named for Joseph Henry). When the inductance and capacitance of a circuit are known, the formula $f = 1/(2\pi\sqrt{LC})$ will give its *resonant frequency*, the frequency at which the circuit will naturally vibrate. A circuit will also emit waves at its resonant frequency. From estimates of inductance and capacitance, Hertz calculated the frequency in his transmitting antenna to be 35.6 million cycles per second, close to the result of 32.3 million cycles per second that he obtained from the formula $f = c/\lambda$.

These numbers were near enough to Maxwell's prediction to satisfy other scientists that electromagnetic waves existed and traveled at the speed of light.[4] The unit of wave frequency is now named *hertz* (one hertz or Hz equals one cycle per second). The electromagnetic waves confirmed by Hertz fell within a range of frequencies that came to be known as the radio frequency spectrum (sidebar 7.2).

Marconi and the Wireless Telegraph

Hertz saw the confirmation of electromagnetic waves only as an advance in scientific understanding, not as a step toward anything useful. Wave reception attenuated with distance, and it appeared that the waves could not be detected beyond about one-half of a mile. Guglielmo Marconi (1874–1937) saw in electromagnetic waves a new way to communicate, however, and he devised a way to transmit them over longer distances (figure 7.1). The son of an Italian landowner and a Scots-Irish mother, Marconi learned of Hertz's achievement from a neighbor, Augusto Righi, a professor of physics at the University of Bologna, who wrote an obituary of Hertz in 1894. Marconi realized that stopping and starting the voltage in a spark transmitter, so as to send out electromagnetic waves in long and short bursts, could emulate the dots and dashes of a telegraph message.

Better reception was possible with a device called a *coherer*, a small glass tube of metal filings placed in the gap of the receiving loop. Electromagnetic waves caused the filings in the tube to align in a way that reduced the resistance of the receiving circuit and strengthened the signal. Marconi developed a coherer that was practical for receiving telegraph signals. He also began to experiment to see if he could transmit waves over distances longer than half a mile. Marconi knew from wired telegraphy that the two ends of a telegraph line could be grounded in the earth, and he wondered if a wireless telegraph could work on the same principle. He redesigned the transmitting antenna to be a vertical line with one end in the ground, rather than the horizontal "dipole" that Hertz had used. Marconi found that he could transmit a signal a mile and more.[5]

Sidebar 7.2   **The Radio Frequency Spectrum**

Electromagnetic waves within a certain range of frequencies are known as radio waves. These are divided into smaller frequency ranges with associated wavelengths. One hertz is a frequency of one cycle per second. One thousand hertz is one kilohertz, one million hertz is one megahertz, and one billion (one thousand million) hertz is one gigahertz. Wavelengths become smaller as frequencies become higher.

Frequency Bands	Frequency Range	Wavelength Range
Very low	3–30 kilohertz	100,000–10,000 meters
Low	30–300 kilohertz	10,000–1,000 meters
Medium	300–3,000 kilohertz	1000–100 meters
High	3–30 megahertz	100–10 meters
Very high	30–300 megahertz	10–1 meters
Ultra high	300–3,000 megahertz	1 meter–10 centimeters
Super high	3–30 gigahertz	10–1 centimeters
Extremely high	30–300 gigahertz	1 centimeter–1 millimeter

Marconi was in no position to compete with wired telegraphy on land, but wireless offered a way to transmit signals over water, especially to ships at sea. Unable to interest the Italian government, he went to England, where he had relatives through his mother. With their support, on June 2, 1896, he filed a provisional patent in London. He demonstrated his invention to British military and postal officials on September 2, when he sent messages—in Morse code—almost two miles across Salisbury plain. A test of nine miles across the Bristol Channel in May 1897 proved the system's value for maritime communication and Marconi received a full patent in July of that year. In March 1899 he transmitted across the English Channel, and in December 1901 he received a message across the North Atlantic from Cornwall to Newfoundland.

A number of British scientists worried that Marconi's patent would place a new technology of great military value under foreign control. They also resented the success of a youthful Italian whose patent challenged the claim by a leading scientist, Sir

Figure 7.1. Guglielmo Marconi. Source: *Cassier's Magazine*, 21:3 (January 1902): 216.

Oliver Lodge, to have demonstrated wireless telegraphy in an 1894 lecture. There is no doubt that Marconi was inspired by Hertz's work, and it is therefore fair to call his system a direct application of science by an engineer. But Lodge did not claim to have invented wireless telegraphy until after Marconi had demonstrated his own invention.[6] The Marconi patent survived challenge, and in 1909 he received the Nobel Prize in Physics in recognition of his work.

Marconi launched a company to provide wireless telegraph services in 1897, and a contract with Lloyds of London, the insurers, gave him a virtual monopoly of wireless communications in British commercial shipping. By 1912 the Marconi company also dominated the wireless business in the United States through a subsidiary.[7] At the height of his success, however, Marconi followed the example of Edison in becoming attached to the system that had brought him success. He took no interest in using radio waves to transmit voice and music. Other inventors soon stepped in to develop wireless telephony and finally radio broadcasting.

Fessenden and the Beginning of Radio Broadcasting

Reginald Aubrey Fessenden (1866–1932) pioneered the transmission and reception of voice by radio (figure 7.2). A Canadian who taught himself electrical technology, Fessenden became a member of Thomas Edison's laboratory staff in 1886 but lost his job after Edison merged his companies in 1889–90. Fessenden soon found work with the Westinghouse Company and joined the faculty of what is now the University of Pittsburgh in 1893 as a professor of electrical engineering. News of Hertz's work fired him with an interest in radio waves, and in 1902 Fessenden organized a company with some investors in Pittsburgh to develop a radio system that would carry voice.[8]

Fessenden connected a telephone to a radio transmitting circuit but found that sparked electromagnetic waves were not practical for carrying voice to headphones at the other end. Sparked waves went out in bursts that could emulate the dots and dashes of Morse code, but the transmissions were a jumble of frequencies that diminished in amplitude. The audio frequency on a wired telephone worked by modulating the amplitude of a constant electric current, and Fessenden realized that he would need a radio wave that was also continuous to carry the modulation of an audio frequency wirelessly. This *carrier* wave, as it was later called, also needed to be emitted at a high frequency (in thousands of cycles per second) for practical reception. A sufficiently powerful alternating-current generator would emit constant-amplitude high-frequency radio waves over long distances, and General Electric agreed to build such an alternator for Fessenden. The company gave the job to a Swedish immigrant engineer, Ernst Alexanderson, who in 1906 delivered a machine capable of 50,000 to 76,000 cycles per second.[9]

To hear sound through headphones, the alternating current in the receiving circuit went through a "rectifier" to convert it into direct current. The carrier frequency was

Figure 7.2. Reginald Fessenden. Courtesy of Brown Brothers, Sterling, Pa.

too high to be heard, leaving the audio signal for the listener to hear. Fessenden invented a liquid rectifier to improve the sound, although later radio operators found solid crystals more practical.[10] On Christmas Day 1906, Fessenden successfully broadcast voice and music from Brant Rock, Massachusetts, to ships offshore. Changing the amplitude of a carrier wave came to be known as *amplitude modulation* or AM (sidebar 7.3). It is today one of the two principal ways that radio stations broadcast (the other way is *frequency modulation* or FM, discussed later in this chapter).[11]

Sidebar 7.3 **Modulation: Amplitude and Frequency**

From Sparked to Continuous Waves

Sparked waves

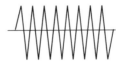

Continuous waves

The first radio signals were produced by spark transmitters. With each change in the direction of an alternating voltage, the spark transmitter emitted an electromagnetic wave. Wireless telegraphy sent out waves in long and short bursts containing a jumble of frequencies of diminishing amplitude. Uninterrupted or continuous waves were necessary for transmitting voice.

Carrier Wave Modulation

Audio Frequency
from Microphone

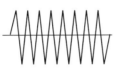

Carrier Wave

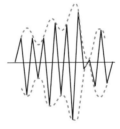

Modulated Carrier

Radio waves with a continuous frequency are called *carrier waves* when they are modulated to carry an audio source. In *amplitude modulation* (above), the audio source modulates the amplitude of the carrier wave. In *frequency modulation* (below right), the audio source modulates the frequency but not the amplitude of the carrier.

Amplitude Modulated (AM) Signal

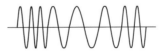

Frequency Modulated (FM) Signal

Source: Abraham Marcus and William Marcus, *Elements of Radio* (New York: Prentice-Hall, 1943), pp. 51–63, 605–18.

Fessenden was on the threshhold of a great breakthrough. But to be practical he needed to go much higher into the radio frequency spectrum (radio stations today broadcast AM signals from 540 to 1,600 kilohertz). American Telephone and Telegraph nevertheless expressed an interest in his work. Although telephone engineers had begun to amplify calls over existing wire lines, they were also open to the possibility of wireless telephony. An agreement to buy Fessenden's patents was derailed, however, in a 1907 reorganization of AT&T. The company did not revive its interest in radio until 1912 and Fessenden's company had gone bankrupt the year before.[12] By then, a better radio system was emerging from the work of other engineers.

The Edison Effect and the de Forest Audion

Transmission was not the only challenge to early radio. Reception was also difficult. Sparked radio waves went out in bursts of frequencies that often interfered with each other. To reduce this interference, Marconi tried to design his transmitters and receivers to operate within a narrower range of frequencies, but sparked transmissions could not be tuned to a single one.[13] A single frequency was possible to transmit in the continuous waves pioneered by Fessenden. A frequency could be captured by a receiving circuit if the circuit was in *resonance*, or in tune, with it. A circuit was in resonance when the inductance (L) and capacitance (C) of the circuit equaled the frequency (f) according to the formula $f = 1/(2\pi\sqrt{LC})$. By varying the inductance or the capacitance, the receiving circuit could be tuned to a particular frequency for reception. It was more practical to fix the inductance and vary the capacitance, and early radio receivers began to have a knob that could be turned to change the capacitance and thus the frequency. Early radios had a receiving or tuning circuit that led through a rectifier into a headphone circuit for listening (sidebar 7.4).[14]

Unfortunately, the audibility of early radio was still very faint. One month before Fessenden's first broadcast, Lee de Forest (1873–1961) designed and built a vacuum bulb that he called an *audion*, which greatly amplified the voice in a wireless receiver (figure 7.3). De Forest attended Yale University as an undergraduate and then graduate student, where he trained under Josiah Willard Gibbs (1839–1903), one of America's most distinguished physicists. The discovery of electromagnetic waves excited

Sidebar 7.4   **Reception and Tuning**

The Tuning Circuit

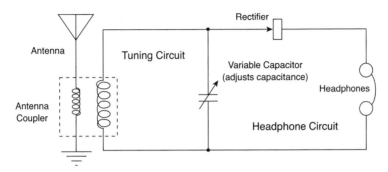

Early radio receivers typically had an antenna, a tuning circuit, a rectifier, and a headphone circuit. An *antenna coupler* transmitted the radio signal from the antenna to the tuning circuit like a transformer (to prevent the resistance in the antenna from adding to the resistance in the tuning circuit and to step up the voltage in the tuning circuit). The alternating voltage went through a rectifier that converted it into a fluctuating direct current. The average of the fluctuating current then sounded the headphones.

Resonance and Frequency

The tuning circuit selected an incoming radio frequency when the circuit resonated at the frequency according to the formula $f = 1/(2\pi\sqrt{LC})$. The circuit could be tuned by varying the inductance or the capacitance. Inductance in a radio receiver was usually fixed, and a listener varied the capacitance to select a frequency for reception.

When $L = 300 \times 10^{-6}$ henries and C is set to 166×10^{-12} farads, then:

$$f = 1/6.28\sqrt{300\times166\times10^{-18}}$$

$$f = 10^9/6.28 \times 224$$

$$f = 710 \text{ kilohertz}$$

The frequency of 710 KHz was the frequency for WOR, an early radio station in New York City. Changing C to 110×10^{-12} farads would select $f = 880$ KHz, the frequency of radio station WCBS in New York City.

Source: Abraham Marcus and William Marcus, *Elements of Radio* (New York: Prentice-Hall, 1943), p. 453.

139

Figure 7.3. Lee de Forest. Courtesy of the New-York Historical Society, New York, NY. Negative no. 49536.

de Forest and he wrote his doctoral dissertation about them. After receiving his Ph.D. in 1899, de Forest tried and failed to get employment with Nikola Tesla and was rebuffed by the Marconi Company. De Forest found investors two years later, though, and in 1902 he launched a company to build spark transmitters and receivers. De Forest's enterprise quickly became the largest radio company in America, with capital raised from an enthusiastic public. After Fessenden won a patent suit against him in 1906, however, de Forest resigned and lost all of his stock.[15]

De Forest patented his audion on January 29, 1907. His invention built on work that had begun in 1883, when Thomas Edison had noticed that the filaments in his light bulbs caused the glass to blacken. Edison inserted a plate between the filament and the glass. When he attached the plate to an electric meter, he was surprised to discover an electric current in the plate, even though no wire connected the plate to the filament. Edison patented the device and did not explore it further.[16] But a

British engineer who had served as a consultant for Edison, John Ambrose Fleming (1849–1945), was intrigued. Electrons were discovered in 1897, and Fleming realized that the effect observed by Edison was the result of electrons ejected by the heat of the filament.

Fleming built a vacuum bulb in 1904 into which he inserted a metal plate a short distance from a filament. A battery supplied positive or negative charge to the plate, and another battery (not on the same circuit) heated the filament. When the plate battery supplied positive charge, the plate attracted electrons from the filament. When the plate was negative, it repelled them (sidebar 7.5). Fleming then replaced the plate battery with an A.C. generator and found that electrons flowed when the current went in a positive direction. He then replaced the generator with a radio tuning circuit and found that his device worked as a rectifier, allowing positive current to flow into the headphones and blocking negative current. The bulb came to be known as a *diode*: a vacuum tube with two elements, the filament and the plate. Fleming's diode was a more reliable rectifier than a crystal; however, it did not improve the audibility of radio.[17]

De Forest realized that with one innovation he could make the diode into an amplifier as well as a rectifier. He disconnected the diode plate from the tuning circuit and instead connected the tuning circuit to a small metal grid that he inserted between the filament and the plate. When voltage in the grid was negative, nothing happened. But when the grid was positive, it attracted electrons from the heated filament and passed them through to the plate, which was positively charged by a battery (sidebar 7.5). The many more electrons that reached the plate now, pulled by the grid, amplified the positive current reaching the headphones. De Forest's device became known as a triode, since it had three elements: the filament, the plate, and the grid.[18] The audion was de Forest's only signficant contribution to radio, and he proved no more successful than Fessenden at bringing his idea to market. He was unable to sell the audion until the American Telephone and Telegraph Company finally purchased his patents in 1913 for $50,000. The telephone company wanted the audion not as a radio amplifier but as a way to amplify long-distance telephone calls.[19]

Sidebar 7.5 **Electronics: Diode and Triode**

The Fleming Diode

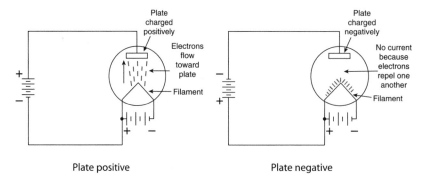

Plate positive   Plate negative

In his diode, J. A. Fleming placed a small plate close to a filament and connected each element to a battery. When the plate was positively charged, relative to the filament, electrons flowed from the filament to the plate. When negatively charged, electrons were repelled from the plate. The diode served in this way as a rectifier: it could take the alternating voltage of an incoming radio signal and convert it into fluctuating direct current that headphones could pick up. Unfortunately, it did not amplify the signal.

The de Forest Audion Triode

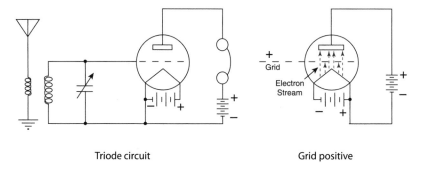

Triode circuit   Grid positive

Lee de Forest inserted a small grid between the plate and the filament in a Fleming diode, attached the grid to the tuning circuit, and detached the plate from it. When the grid was positively charged by the alternating voltage of the signal, it pulled electrons from the filament to the positively charged plate. When the grid was negative, the electrons did not go through. De Forest's triode amplified the electron flow to the plate and thus amplified the signal reaching the headphones.

Source: Abraham Marcus and William Marcus, *Elements of Radio* (New York: Prentice-Hall, 1943), pp. 97–119.

Armstrong and the Regenerative Circuit

Fessenden and de Forest made fundamental contributions to radio, but they did not create commercially viable systems for transmitting and receiving voice and music. To succeed as a broadcast medium, radio needed the work of a third figure, Edwin Howard Armstrong (1890–1954), whose regenerative circuit and superheterodyne receiver finally made commercial broadcasting practical. As a high school student, Armstrong had built a sophisticated wireless receiving station at his home in Yonkers, New York. In the fall of 1909, he entered Columbia University, where he studied under Michael Pupin, who had co-invented inductive loading to amplify telephone calls (see chapter 3) and who went on to conduct research in radio.[20]

Armstrong

In his junior year at Columbia, Armstrong discovered a way to improve radio reception in a triode receiver. In early radios, the receiving antenna passed alternating voltage to the tuning circuit through an *antenna coupler*. In the coupler, a coiled section of the antenna and a coiled section of the tuning circuit were placed close together. Current from the antenna induced voltage in the tuning circuit without adding the antenna's resistance to that of the tuning circuit.[21] Armstrong's innovation was to coil a section of the plate circuit and position it near the antenna coupler (sidebar 7.6). The current in the plate circuit fed back to the tuning circuit and made the radio signal stronger. Armstrong, who termed his invention a *regenerative circuit*, in this way made the triode not only a rectifier and an amplifier but a stronger amplifier.[22] One problem with Armstrong's innovation was that too much regeneration caused the tuning circuit to emit radio waves as well as receive them. These waves interfered with the incoming signal and caused a howl in the headphones. Armstrong found that by moving the coiled section of the plate circuit farther away from the antenna coupler, the feedback could be held just below the level at which the howl occurred.[23]

Armstrong failed to interest the Marconi Company or American Telephone and Telegraph in his regenerative circuit. He finally licensed the idea to Telefunken (now AEG Telefunken), the leading radio company in Germany. His relationship with the Germans ceased when the United States went to war with them in April 1917. Armstrong joined the Army as a captain and served in France from 1917 to 1919 (figure 7.4), where he helped improve wireless communications for the army and its airplanes.

143

Sidebar 7.6   **Armstrong and Radio**

The Regenerative Circuit

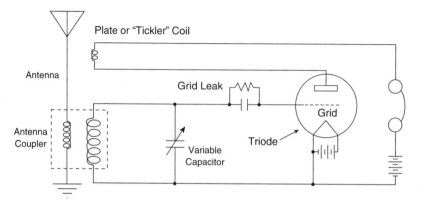

Edwin Howard Armstrong strengthened the amplification in radio receivers by feeding the plate current back into the tuning circuit. He coiled a section of the plate circuit (called the *tickler coil*) and positioned it close to the tuning coil in the antenna coupler. The plate current fed back into the tuning circuit, strengthening the radio signal. Armstrong kept the strength just below the level at which the tuning circuit and triode would emit radio waves themselves. A grid leak siphoned excess electrons from the grid.

The Superheterodyne Receiver

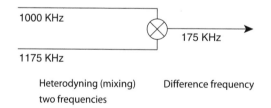

Amrstrong's "superheterodyne" receiver included a circuit that could generate a frequency a fixed distance from an incoming radio signal. Changing this local frequency tuned the radio. Mixing the incoming frequency with the local one produced a difference frequency that the radio detected and amplified for listening.

Source: Abraham Marcus and William Marcus, *Elements of Radio* (New York: Prentice-Hall, 1943), pp. 121–29, 231–47.

Figure 7.4. Edwin Howard Armstrong. Courtesy of the Rare Book and Manuscript Library, Columbia University, New York, NY. Armstrong Papers Collection.

While in France, he solved another fundamental problem with radio technology by finding a way to use an idea advanced by Reginald Fessenden.

In addition to his innovation of amplitude modulation, Fessenden had proposed a way to improve reception by *heterodyning* frequencies. It had long been known that when two musical tones are played together, they created a third tone with a frequency equal to the difference between the first two. Fessenden had reasoned that two closely spaced high-frequency radio waves would do the same and produce a much lower *difference* (or *intermediate*) frequency.[24] The difference frequency would be the same shape as the modulated carrier wave but would be easier to tune. Fessenden did not have a practical way to create a second high frequency in a receiver, though, and transmitting one by alternator was expensive. Armstrong realized that the ability of a regenerative circuit to generate as well as amplify a signal could make heterodyning practical. With a second triode, Armstrong devised a circuit that created a second high frequency in the receiver to heterodyne against an incoming radio wave. The innovation sharpened reception and permitted a smaller receiving antenna. Armstrong called the new radio a "superheterodyne" receiver and modern AM radios are of this type.[25]

With the regenerative circuit and the superheterodyne receiver, radio finally had the audibility and selectivity to be practical for a mass market. In 1918 the Institute of Radio Engineers in the United States gave Armstrong its Medal of Honor for inventing the regenerative circuit.[26] At the end of the summer of 1919, he returned home. He soon faced challenges to his major patents as well as opportunities to play a central role in a completely new organization devoted to radio.

Sarnoff and RCA

As a wartime measure, the U.S. Navy took control over all radio communication in the United States in 1918. After hostilities ended, the future of American radio was uncertain. Lieutenant Commander (later Admiral) Stanford C. Hooper, in charge of radio for the Navy, did not want the nascent radio industry to return to the American Marconi Company, the subsidiary of a foreign firm, but neither did he favor continued government control. After negotiations brokered by Owen D. Young, vice president of General Electric, the Navy released the radio system to a new American corporation controlled by GE. The new firm, the Radio Corporation of America (RCA), incorporated on October 17, 1919, with Young as chairman and Edward J. Nally, formerly vice president and general manager of American Marconi, as president. The firm absorbed the prewar American Marconi network.[27]

The future direction of RCA was soon set, however, by an assistant to Nally, David Sarnoff (1891–1971). Born in Minsk, Russia, Sarnoff came to the United States in 1900 and went to work in 1906 as a telegraph messenger. Fired when he refused to work on the Jewish holy days of Rosh Hashanah and Yom Kippur, he found another job as an office boy for the American Marconi Company and taught himself radio technology by reading in the company library, visiting installations, and talking to engineers, including Marconi, to whom he introduced himself at the age of fifteen. Sarnoff worked as a radio telegrapher (figure 7.5), and rose to became commercial manager of the American Marconi Company in 1917. He joined RCA with Nally.[28]

The new corporation did not control the key U.S. radio patents. American Telephone and Telegraph owned the de Forest audion, and Westinghouse in 1920 had purchased Armstrong's patents for the regenerative circuit and the superheterodyne receiver and also possessed the Fessenden patents. On November 4, 1920, Westinghouse began its own radio broadcasting station, KDKA in Pittsburgh, and other stations began in Newark and Chicago. Owen Young asked Sarnoff to assess the future of RCA, and the young man responded with a twenty-eight-page memorandum on January 31, 1920. Sarnoff urged Young to acquire the major radio patents, make and sell radio sets to the public, and operate a network of radio stations to broadcast news and entertainment. Not seeing as large an audience for radio as Sarnoff, AT&T and Westinghouse agreed to cross-license their patents with those of RCA and not compete with RCA in radio broadcasting. RCA gave AT&T 10.3 percent and Westinghouse 20.6 percent of its stock in return.[29]

Society played an important role in shaping the use of radio technology. Marconi's original idea was to imitate telegraphy through wireless communication from one point to another. Fessenden and later radio entrepreneurs saw a larger market for broadcasting from one point to many, and in the 1920s audiences responded. With technology licensed from RCA, stations began broadcasting in cities and towns across America. In 1926 RCA created the first network of stations, the National Broadcasting Company (NBC). A rival radio network, the Columbia Broadcasting System (CBS), grew alongside Sarnoff's under another young Russian immigrant, William S. Paley. The first programs covered sports events but quickly broadened to include news, music, and other forms of entertainment (figure 7.6).[30] RCA manufactured and sold radio receivers and branched into producing phonograph players and recordings. Triode vacuum tubes could produce high frequencies more efficiently than alternators and soon took their place in the transmission of radio waves.[31]

Figure 7.5. David Sarnoff as a radio telegrapher in 1908. Courtesy of the David Sarnoff Library, Princeton, NJ.

The new industry also needed regulation as much as it needed an audience. Spark transmitters disrupted the continuous waves needed for broadcasting and were made illegal in 1923. The Radio Act of 1927 created the Federal Radio Commission, which became the Federal Communications Commission in 1934. The commission established the public right to license broadcasters to operate at assigned frequencies and levels of station power.[32] RCA came under attack for its concentration of patents and

Figure 7.6. John and Kenneth Coolbaugh (uncles of senior author) with radio set. Source: Billington family album.

control over the radio market. The Federal Trade Commission began an investigation in 1922 to see if the firm violated the Sherman Antitrust Act by uniting the largest electrical companies. Under a consent decree in 1932, General Electric and Westinghouse divested themselves of ownership, AT&T having done so earlier.[33] Sarnoff took over RCA in 1930 and served as its president until 1965.

Patents and Priorities

During the 1930s, RCA became entangled in a conflict with Edwin Howard Armstrong. David Sarnoff had met Armstrong in Michael Pupin's laboratory in 1913, and the two had become friends. In the early 1920s, Armstrong advised RCA on technical matters, and the company purchased some of his patents, making him wealthy.[34] But a lengthy dispute between Armstrong and de Forest over who invented the regenerative circuit eventually drove Armstrong and Sarnoff apart.

Back in March 1915, in a report to the Institute of Radio Engineers, Armstrong noted that residual gases in the audion tube did not play a role in its working, directly contradicting the explanation that de Forest had given of how his audion worked.[35] De Forest then claimed that he himself had discovered the regenerative circuit several years earlier. One of de Forest's assistants had made a vague sketch of a circuit resembling Armstrong's in 1912, although de Forest did not realize that it could be regenerative until Armstrong had showed what such a circuit could do. In 1915 de Forest filed for patent interference, a kind of claim in which he argued that Armstrong's patent was based on an idea that he, de Forest, had conceived first. On his return from France Armstrong countersued de Forest. Experienced in radio suits, federal Judge Julius M. Mayer of the Southern District of New York decided the case in Armstrong's favor on May 17, 1921. De Forest appealed and on March 13, 1922, the U.S. Court of Appeals in New York ruled unanimously against de Forest. The Patent Office also ruled against him.

Armstrong had won a big victory, which included having de Forest pay his legal fees. But de Forest was practically insolvent. Rather than forgo a penalty he could not collect, Armstrong stubbornly insisted on payment. De Forest then challenged the Patent Office decision in the U.S. Circuit Court for the District of Columbia. On May 8, 1924, Judge Josiah van Orsdel gave de Forest the rights to regeneration. More legal defeats followed for Armstrong, who finally took the case to the U.S. Supreme Court. On October 29, 1928, without addressing the technical merits of the case, the high court gave the invention to de Forest. Despondent after losing his legal battle, Armstrong tried to return the medal awarded to him by the Institute of Radio Engineers in 1918. The institute refused to take it back and reaffirmed its view that Armstrong was the true inventor of the regenerative circuit.[36]

By the early 1930s, RCA owned the Armstrong patent and the rights to the de Forest patents as well. Sarnoff had to choose which inventor to back and he made a business decision that led to the end of his friendship with Armstrong (figures 7.7 and 7.8). De Forest's claim to the regenerative circuit dated from the court ruling in 1924; if upheld,

Figure 7.7. David Sarnoff in later life. Courtesy of the David Sarnoff Library, Princeton, NJ.

RCA could collect royalties on it until 1941. Armstrong's patent dated from 1914 and its patent rights ran out in 1931: Sarnoff sided with de Forest.[37] In retaliation, Armstrong bought a controlling interest in a small company that infringed de Forest's claim. Sarnoff and RCA filed suit in a case that went up to the U.S. Supreme Court. On May 21, 1934, Justice Benjamin Cardozo delivered the majority opinion against Armstrong with scientific justifications that revealed a serious misunderstanding of modern electronics.[38] Armstrong also lost his claim to the superheterodyne receiver to a French rival, whose claim was upheld by U.S. courts despite the rival's failure to produce a working system.[39] These cases made clear that in a more technologically complex world, engineering literacy among decision makers was an urgent need.

Figure 7.8. Edwin Howard Armstrong in later life. Courtesy of the Still Picture Branch, National Archives, College Park, MD. RG173-RAD-19-1.

Radio versus Television

Despite his legal difficulties, Armstrong continued to seek improvements in radio. One great technical challenge remained: the problem of static. Radio static is a spiking in the amplitude of a radio wave. It is especially noticeable during electrical storms and when radio signals are weak. Armstrong began to study this problem as a student, and his last contribution to radio was to develop *frequency modulation* or FM as a way to overcome this problem. In FM the amplitude of the radio waves remained constant (sidebar 7.3). The audio signal was instead fed into the radio signal in a way that varied the frequency of the radio waves rather than the amplitude. FM virtually eliminated the problem of static.[40]

Armstrong took out five patents on FM between 1930 and 1933. In December 1933, using an RCA antenna later placed atop the newly completed Empire State Building, he demonstrated to David Sarnoff the high fidelity possible with FM transmission. But in July 1935 RCA ordered him to remove his equipment. For Sarnoff, there were two problems with allowing FM radio to develop. First, RCA had a highly profitable business tied exclusively to AM radio. Second, and more important, research had opened the possibility of sending television transmissions using electromagnetic waves. Sarnoff believed that television would someday replace voice-carrying radio, just as radio had replaced wireless telegraphy.[41]

Armstrong moved across the Hudson River to New Jersey, constructed a tower, and began to design a system for FM broadcasting. He was in business by July 1939. General Electric began manufacturing FM receivers, and some stations in New England began to broadcast in FM. Realizing that it could not stop FM, in June 1940 RCA offered Armstrong $1 million for a license. Armstrong's patent lawyer was dumbfounded when his client flatly turned down the offer. But Armstrong wanted 2 percent of all the earnings on FM receivers and equipment, a condition that General Electric and other manufacturers had accepted. Sarnoff did not accept. RCA infringed the Armstrong FM patents, and Armstrong sued in 1948.

Armstrong's patents would run out in 1957 and Sarnoff adopted a strategy of delay, draining his rival of funds. Armstrong rejected an RCA offer to settle for $200,000, and on the evening of January 31, 1954, Edwin Howard Armstrong climbed out of his thirteenth-storey apartment window in Manhattan and fell to his death below. After-ward, his widow settled with RCA and later won other FM cases that had plagued her husband.[42] By then, too, the radio rivalry had resolved itself, although not without irony. AM and FM developed alongside each other, each finding a unique niche. And radio itself did not die out with the coming of television but continued alongside the new medium and remains a thriving technology today.

Radio was a case of applied science, in which a scientific discovery by Hertz led quickly to a major new technology that Marconi designed for wireless communication. Fleming was stimulated to invent his diode by the discovery of electrons in 1897, but he missed the engineering insight that led de Forest to add a grid and create a workable vacuum tube amplifier, the triode. Armstrong was more typical of later trained engi-neers, but his improvements to radio were of fundamental importance. Modern elec-tronics began with the use of electron flows in vacuum tubes to amplify telephone calls and radio transmissions. After 1948 transistors and then microchips would take the place of vacuum tubes in electronic circuits. But predictions that radio would be re-placed were not fulfilled, and radio remains an important branch of telecommunica-tions today.

CHAPTER EIGHT

Ammann and the George Washington Bridge

Amore closely interconnected civilization emerged from the new networks, processes, and machines that spread in the early twentieth century. New bridge structures supplied a different form of connection no less vital. By 1939 the bridges of New York, Philadelphia, Washington, St. Louis, and San Francisco were essential to the everyday life of those cities and their surrounding areas. The need for new and larger bridges was a natural outgrowth of the demand for more and better roads in the 1910s and 1920s. The availability of inexpensive steel in the late nineteenth century gave civil engineers a new material with which to design and build bridges of much greater size and strength than traditional structures of wood and stone.

Steel bridge engineering made a spectacular advance in the George Washington Bridge, completed in 1931 over the Hudson River between New Jersey and New York. Designed by Othmar Ammann, the structure was almost twice the span of any previous bridge and became a model for the design of other long-span bridges. The structure was innovative in two ways. First, Ammann employed a new and daring theory to estimate the weight of traffic that achieved dramatic economy in the cost of the bridge. Second, he relied on a new theory that justified minimizing the vertical depth of the roadway deck. This theory, while successful, led to danger when engineers relied on it for lighter suspension bridges later in the 1930s.

Bridges from Iron to Steel

As a bridge-building material, iron was far stronger than wood and far less costly than stone. Modern techniques improved the quality and quantity of iron and gave structural engineers the freedom to bridge longer spans. The first great designer of iron

155

bridges, Thomas Telford (1757–1834), built a graceful arch of cast iron in his 1814 Craigellachie Bridge in Scotland. Telford's Menai Straits Bridge in Wales, completed in 1826, used wrought iron chain-link cables to suspend a roadway of unprecedented length, 580 feet between its two towers.[1] With maintenance, these bridges proved durable and economical. Telford's two bridges along with many others that he built are still in use.

The coming of the railroad in the mid-nineteenth century created a demand for iron bridges of greater size and strength. The French engineer Gustave Eiffel (1832–1923) built wrought iron arch bridges, such as his 541-foot Garabit Viaduct in southern France, to carry trains around mountains and over gorges. The pylon towers supporting his elevated railway at Rouzat anticipated the design of Eiffel's famous 1889 tower in Paris.[2] John A. Roebling (1806–69) built a number of iron suspension bridges in the United States, culminating in his 1866 Cincinnati Bridge, which spanned 1,057 feet.[3] With the development of structural steel in the 1850s and 1860s, civil engineers designed bridges that could span longer distances and carry heavier loads. James B. Eads (1820–89) completed the first steel arch bridge in 1874 spanning the Mississippi River at St. Louis.[4] John Roebling's Brooklyn Bridge, finished by his son and daughter-in-law in 1883, was the first suspension bridge in steel. The Brooklyn Bridge used treated (galvanized) steel wire to prevent rust and achieved a record span of 1,595 feet.[5]

A bridge has to carry two kinds of weight (sidebar 8.1). Its *dead load* is the static weight of the structure itself. Its *live load* is mainly the weight of traffic crossing it. Live loads also include any additional dynamic pressures acting on the structure, such as wind. Arch bridges work in *compression*: under the roadway, an arch supports the weight of the bridge and the traffic going across it. An overhead arch also works in compression when it suspends a roadway below it. Cable bridges work in *tension*: vertical wires suspend a roadway deck from cables, which are strung over towers and anchored on each side of the bridge. In an arch, the two abutments or end points have to resist the weight of the bridge in two directions, horizontally and vertically. A cable has to resist the same forces, the vertical being greatest at the tower tops.

Sidebar 8.1   **Bridges: Forces and Reactions**

A bridge must carry its own weight, called *dead load*, and the weight of traffic going across, or *live load*.

An arch bridge can support a roadway with arches from below or from above. The arch carries loads by *compression*: the weight on the arch is transmitted to the two abutments (the two end points) as a force pressing down vertically and as another force pushing out horizontally.

The vertical force (V) is calculated at the abutments and is the product of the load (q) and the length or span of the arch (L) divided by two. The division by two reflects the fact that one-half of the total weight goes to each abutment.

The horizontal force (H) is calculated at the abutments and also at the midspan or crown of the arch. The force is equal to the product of the load per unit of length (q) and the total length (L) squared, divided by the vertical rise of the arch (d) times eight.

The two forces are resisted at the abutments by horizontal and vertical *reactions* that are calculated by the same formulas.

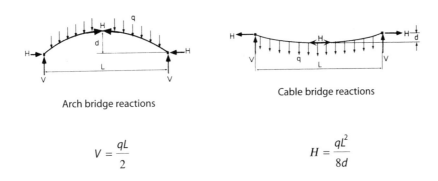

Arch bridge reactions

Cable bridge reactions

$$V = \frac{qL}{2}$$

$$H = \frac{qL^2}{8d}$$

A cable suspension bridge works in *tension*: wires suspend the roadway deck from cables that go over towers and are anchored to the abutments. The pulling forces in the cable are also vertical and horizontal and are the same as for an arch bridge. The forces and reactions in the cable are calculated from the tower tops rather than from the abutments, and *d* is the cable sag rather than the rise of the arch.

Source: David P. Billington, *The Innovators: The Engineering Pioneers Who Made America Modern* (New York: John Wiley and Sons, 1996), p. 8.

Bridging the Hudson

The Brooklyn Bridge connected Manhattan to Brooklyn and facilitated the union of the five boroughs of New York City in 1898. By then, New York was drawing commuters from a wide area, including Long Island, southern Connecticut, southern New York State, and northern New Jersey. The city needed to be supplied from even greater distances. Steel bridges eventually connected Manhattan to the boroughs of the Bronx and Queens, and linked Staten Island to New Jersey. But New York's greatest need was to cross the Hudson River between Manhattan and New Jersey.[6]

Plans to bridge the Hudson began soon after the Civil War and the states of New York and New Jersey jointly authorized a bridge in 1890. Financing did not materialize, though, and after 1900 the Pennsylvania Railroad built tunnels under the river instead. But the growth of motor vehicle traffic following the introduction of the Model T and other cars made new road crossings urgent. Ferry service could not accommodate the volume of traffic, and interest in a great bridge over the Hudson River revived after the end of World War I in 1918.[7]

In 1888 the engineer Gustav Lindenthal (1850–1935) had proposed a suspension bridge to carry railway traffic over the Hudson from mid-Manhattan to New Jersey. As bridge commissioner of New York in 1902–3, he initiated several great bridge projects to connect the boroughs of the city. After these, his most important project was the Hell Gate Bridge between Ward's Island and the Borough of Queens. The 1917 Hell Gate was the longest-spanning steel arch bridge in the world and was designed to carry a railroad. In 1920 Lindenthal revived and expanded his earlier proposal for a Hudson River bridge near 57th Street in Manhattan.[8]

Lindenthal's 1920 design (sidebar 8.2) called for a steel suspension bridge. A single span 150 feet above the water at its center would extend 3,240 feet in length between two giant towers on the river embankments. The design was dramatic and would have spanned a distance twice as long as the Brooklyn Bridge. But Lindenthal believed that the principal use of his bridge would be to carry railway traffic, and he designed the bridge to carry the enormous weight of twelve fully loaded freight trains as well as four lighter rapid transit lines and sixteen lanes of motor vehicle traffic. The estimated cost of his bridge, $180 million, was far higher than the $11 million that a motor vehicle

Sidebar 8.2 **Railroad versus Automobile**

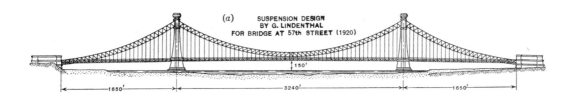

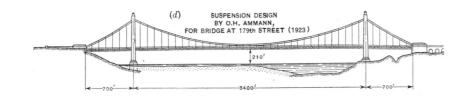

In 1920 Gustav Lindenthal designed a bridge over the Hudson River at 57th Street in Manhattan to carry twelve freight trains on a lower deck and sixteen lanes of motor vehicle traffic, four transit rail lines, and pedestrians on an upper deck. His bridge would have cost $180 million.

Othmar Ammann proposed a bridge in 1923 to cross the Hudson at 179th Street to carry two pedestrian walkways and eight lanes of motor vehicle traffic on an upper deck and four transit rail lines on two lower decks. He estimated the bridge and its approaches to cost $30 million.

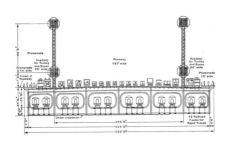

Cross section of Lindenthal deck

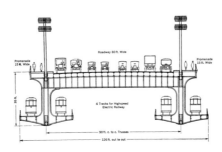

Cross section of Ammann deck

Sources: O. H. Ammann, "General Conception and Development of Design," *ASCE Transactions* 97 (1933): 9, 15, 36 (diagrams); 10, 61–62 (costs).

tunnel was estimated in 1913 to cost.[9] In 1919 the states of New York and New Jersey authorized a vehicular tunnel between Manhattan and New Jersey. The Holland Tunnel, named for its designer Clifford Holland on its completion in 1927, proved difficult to construct, and with only two lanes it was not the answer to the need for a major new Hudson River crossing.

Ammann as Public Entrepreneur

Lindenthal's chief assistant on the Hell Gate Bridge, Othmar Ammann (1879–1965), realized that motor vehicles were replacing railroads as the mainstay of twentieth-century transportation. Ammann came to the United States from Switzerland in 1904, two years after graduating from the Federal Technical Institute in Zürich (figure 8.1). He had learned structural design in Zürich from Professor Wilhelm Ritter (1847–1906), who had made the analysis of loads on suspension bridges more rigorous. Ritter also taught that structures were not simply utilitarian structures, but could carry their weight by means of efficient and elegant form. Ritter had visited the United States in 1893 to study its bridges, and two years later he published a book on them, the principles of which he taught to Ammann and to Robert Maillart, who would become a master designer in reinforced concrete (chapter 9). His students were impressed by widespread bridge-building underway in the United States.

Ritter's book praised the 1889 Washington Bridge over the Harlem River in New York City as an elegant example of an arch bridge, and he closed his text with a drawing of Gustav Lindenthal's 1888 proposal for a huge suspension bridge over the Hudson River.[10] The opportunity to design and build new bridge structures stimulated the young Ammann to travel to the United States and he eventually settled in the New York area.

Ammann became familiar with Lindenthal's newer design for a Hudson River crossing while working under him. The younger man saw that New York would need crossings primarily for automobiles and trucks, and he believed that a cable suspension bridge to carry motor vehicles and light rapid transit lines could be built at far less cost than Lindenthal's structure (sidebar 8.2). By not carrying heavy railroad trains, such

Figure 8.1. Othmar Ammann in 1930. Source: Courtesy of Dr. Margot Ammann Durer.

a bridge would have no need to connect to the freight terminals of midtown Manhattan. The bridge could be located at the northern end of the island, where the inflow of bridge traffic and the approaches to the bridge would also be less disruptive to the city.

Ammann tried to persuade Lindenthal to adopt a less costly bridge over the Hudson to carry mostly motor vehicle traffic. Lindenthal was unwilling to abandon his vision of a great railroad bridge, but bankers in New York refused to finance it, and Manhattanites worried that traffic from a massive bridge at 57th Street would worsen midtown congestion. After finishing the Hell Gate Bridge in 1917, Lindenthal's office lacked work, so, at his suggestion, Ammann occupied himself for three years managing a clay factory in Trenton, New Jersey, of which Lindenthal was a director along with an attorney, George Silzer. In 1922 Silzer won election to a three-year term as governor of New Jersey, serving from 1923 to 1926. An activist Wilsonian Democrat, Silzer wanted to improve the highway system in New Jersey and build bridges to New York City.[11]

Ammann knew that a great bridge was necessarily a public concern and that political support would be crucial to achieving it. He approached Silzer after his election and won the governor's support for a great motor vehicle bridge over the Hudson from Fort Lee, New Jersey, to 179th Street in Manhattan. Ammann resigned from Lindenthal's office in March 1923 to pursue his vision. He and the governor faced an obstacle, however, in the fact that Fort Lee was located in Bergen County, New Jersey, a Republican stronghold. A great bridge backed by a Democratic governor could encounter political opposition in the county most likely to benefit from the bridge. In his January 1923 inaugural address, Silzer called for a bridge across the Hudson from New Jersey to New York, but over the next year Silzer did not actively pursue the idea. Ammann took the lead in a campaign to generate public support for the bridge. He became an entrepreneur whose goal was to create a public rather than private enterprise.[12]

Local business interests on both sides of the river, including the major daily newspaper in Bergen County, the *Bergen Record*, favored building a tunnel to connect New Jersey and northern Manhattan. Over the summer and fall of 1923, local fund raising for a tunnel began, as Ammann was unable to persuade local leaders on either side of the river to support the building of a bridge. In November 1923, however, the Port Authority of New York (as it was then known) announced a public hearing for December 5 to hear proposals for additional tunnels to accommodate motor vehicles. Formed in 1921 by the governors of New York and New Jersey as a bi-state planning agency, the Port Authority's mandate was to relieve railway congestion in the New York metropolitan area. When the agency proved unable to develop a workable plan for railways, Silzer and Ammann urged the Port Authority to consider new motor vehicle bridges at its hearing, and the agency's board of commissioners agreed.[13]

In the month prior to the December hearing, Ammann prepared the case for a span across the Hudson, estimating that a motor vehicle bridge could be built for $30 million, or $25 million without transit rail lines. He argued that a bridge could support itself through tolls. At the hearing, most speakers agreed on the need for more vehicular crossings. While some supported tunnels, others supported a bridge on the northern end of Manhattan. Dwight Morrow, a partner in the J. P. Morgan Bank of New York, informed Governor Silzer that private funding for a bridge was unlikely but that the Port Authority might raise the money itself. Silzer and Ammann pressed the agency to construct and operate its own bridges and tunnels. On December 17, Ammann submitted a report to Silzer. The governor sent it to the Port Authority staff, and on December 21, the commissioners agreed to carry out a study of a great bridge between 179th Street and Fort Lee.

During the winter and spring of 1924, Ammann held meetings with civic groups in Bergen, Passaic, and Morris counties in New Jersey; in Washington Heights, Harlem, the Bronx, Westchester, and Yonkers in New York; and in southern Connecticut. Success finally began to come when a state senator from Bergen County backed the bridge that spring. The New Jersey and New York state legislatures approved the Port Authority's decision to conduct a study and authorized two bridges to Staten Island. Governor Al Smith of New York, like Silzer a Democrat, favored public control of bridges and tunnels and supported a more active role by the Port Authority. The Harlem Board of Commerce then switched its support from a new tunnel to the Ammann bridge. Finally, the New Jersey State Senate passed a bill in January 1925 authorizing a bridge over the Hudson River at Fort Lee, and the state assembly agreed. The New York state legislature concurred in March.[14]

Ammann submitted a bid for the contract to design the two Staten Island bridges. He lost to the more experienced private bridge design firm of J.A.L. Waddell. But the Port Authority soon began a tradition of relying on its own engineers for design work, and Ammann offered himself as a staff engineer. With Silzer's backing, the Port Authority hired Ammann and entrusted him with the design and supervision of the construction of the great Hudson River bridge.

Ammann as Engineering Designer

Since the early nineteenth century, the U.S. Army Corps of Engineers has regulated navigable rivers in the United States. The Corps of Engineers required a bridge across the Hudson to have a clearance of at least 200 feet above water at its midspan and 185 feet above water near the towers.[15] Army engineers and private navigation interests opposed the use of piers in the middle of the river. For his bridge, Ammann envisioned a single long span with a midpoint 210 feet above the water and a length of 3,500 feet between two towers, one near each embankment. To build such a bridge at a substantially lower cost than Lindenthal's structure, however, Ammann had to rethink some fundamentals of bridge design.[16]

Ammann first questioned the assumptions of what traffic loads a motor vehicle bridge would carry. A theoretical maximum would fill every lane with heavily loaded trucks traveling end to end. This loading would resemble railroad trains in the weight they would exert on a bridge. But in fact this kind of end-to-end loading of heavy trucks was unlikely. Cars and trucks usually traveled with one or two vehicle lengths of space between them and there were usually more cars than trucks. The engineer J.A.L. Waddell had argued in 1916 that for motor vehicle bridges, long spans of more than 800 feet should be designed to support one-half the weight of heavily loaded trucks laid end to end.[17]

Ammann knew that the estimated live load would have a large influence on the amount of steel the bridge would need; and the amount of steel was crucial to the cost of the bridge. He designed the George Washington Bridge to have an eighty-foot wide roadway with eight lanes for motor vehicles, which he believed would need to support 20,000 pounds of live load per foot of span length. Two pedestrian walkways would need to carry 2,000 pounds per foot. On two narrower parallel decks underneath, he designed four lanes for rapid transit trains with a capacity of 24,000 pounds per foot. Adding all the traffic gave a total live load of 46,000 pounds per foot. Taking Waddell's safety factor of one-half suggested a design for 23,000 pounds per foot of span length.

The Delaware River Bridge (now the Ben Franklin Bridge) in Philadelphia was designed to carry one-half of its estimated total live load. But engineers knew that as the span length of a bridge increased, the heaviest vehicles decreased as a percentage of the overall traffic, and a traffic analysis showed that the 1,750-foot span of the Philadelphia bridge could have been designed for one-third, not one-half, of the maximum live load.[18] Ammann decided that the percentage of heavy trucks would be even less on his own much longer span of 3,500 feet. He concluded that the bridge could be designed for an even greater reduction of the maximum load in any one of the lanes reserved for motor vehicles and rapid transit trains.

Ammann developed an estimate of the traffic load from several simple formulas that he devised (sidebar 8.3). He calculated that a span of 3,500 feet for an eight-lane bridge could be designed for about 17 percent of the maximum live load, or about 8,000 instead of 46,000 pounds per foot.[19] Seventeen percent was a dramatic reduction in the estimated live load for a bridge, and it was a stunning reduction for a bridge span that would be twice as long as any in existence. Few drivers who cross the George Washington Bridge today are aware of the fact that the bridge is only designed to carry about one-sixth of its estimated maximum traffic weight. But traffic studies in the 1920s confirmed Ammann's safety factor, which became standard practice.[20] The steel used in the George Washington Bridge represented about $23 million of the bridge's final total cost of $57 million. Had the bridge been designed to withstand one-half of its maximum traffic load, the steel would have likely cost an additional $7.4 million.[21]

To complete his safety estimate, Ammann also calculated the stresses in the steel cables (sidebar 8.4). Steel reaches a *yield point* under stress, at which it deforms before reaching its breaking point. For the steel wire that he planned to use in his cables, the breaking point was about 220,000 pounds per square inch (psi) of stress and the yield point was about 150,000 psi. Engineers normally allow a stress of about one-half the yield point in steel, and Ammann took this *allowable stress* (f_a) in the cables to be 82,000 psi. From this number, he calculated the cross-sectional area needed in the steel cables to be 3,200 square inches. If he had instead calculated the stress under the full maximum traffic load of 46,000 pounds per foot, the allowable stress in 3,200 square inches of cable would have been 148,000 psi. Under this maximum load, the bridge still would have held, although with almost no margin of safety. Responsible public officials would never have tolerated a safety factor so thin, but they accepted Ammann's factor of 150,000/82,000 psi as a reasonable compromise. Technical needs thus depended on a social choice to balance safety and economy.[22]

Ammann's traffic-load estimate was daring but successful. His embrace of a new design theory, however, proved dangerous. The deck of a suspension bridge hangs from a flexible cable. To prevent excessive motion of the deck, engineers usually strengthen the deck of a suspension bridge with trusswork underneath. But a new idea in bridge design called *deflection theory* argued mathematically that the stiffer the cable was relative to the deck, the less would be the effect of live load on the deck. The theory per-

Sidebar 8.3 **Ammann's Traffic Load Estimate**

Before the George Washington Bridge, structural engineers estimated the maximum traffic load on a bridge by imagining what it would weigh with every space filled by heavily loaded trucks laid end-to-end. For the George Washington Bridge, this maximum live load would have been 46,000 pounds per foot of span length. Engineers recognized, however, that motor vehicles usually traveled with one or two vehicle lengths between each vehicle and that more vehicles would be cars than trucks. Engineers before Ammann designed bridges on the assumption that a long-span bridge would need to support 50 percent of its estimated maximum traffic load.

Ammann realized that as the span length of a bridge increased, the likelihood of maximum traffic load would diminish. To calculate this reduction, he invented a formula $K = 0.2 + 160/(200 + l)$, to calculate the percentage of maximum load in one lane (K) on a bridge. For a span l of 3,000 feet, he estimated K to be 0.25 or 25 percent:

$$K = 0.2 + 160/(200 + l)$$

$$K = 0.2 + 160/(200 + 3,000)$$

$$K = 0.2 + 0.04 = 0.25$$

The more lanes the bridge had, the less likely it was that all would carry the same load per foot at the same time. Ammann calculated this likelihood as a percentage by a second formula, $C = 0.5 + 2/(n + 3)$, in which n was the number of lanes on the bridge. For one lane, C was 1.0; for eight lanes ($n = 8$), it was 0.6816, which he rounded to 0.682:

$$C = 0.5 + 2/(n + 3)$$

$$C = 0.5 + 0.182 = 0.682$$

Multiplying the maximum traffic load of 46,000 pounds per foot by 0.682, and then multiplying the result by the 0.25 estimated for one lane, gave a figure of 7,843 pounds per foot, or about 17 percent of the maximum. This was the amount of live load for which Ammann believed the bridge would need to be designed. For simplicity, he rounded this figure up to 8,000 pounds. For the full 3,500 foot span of the George Washington Bridge, K was 0.24 and with C of 0.6818, the live load would have been 7,527 pounds or 16 percent, so Ammann's estimate was slightly more conservative.

Source: Allston Dana, Aksel Anderson, and George M. Rapp, "George Washington Bridge: Design of Superstructure," *ASCE Transactions* 97 (1933): 103.

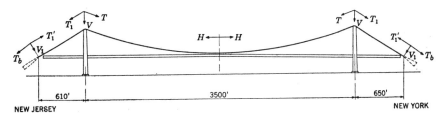

Ammann calculated the total load for the George Washington Bridge by adding his estimate of the traffic or live load to the estimated dead load. He then computed the horizontal force (H) in the cable at midspan (equal to the horizontal force at the tower tops), and the vertical force (V) at the towers, to measure the loads on the main span between the towers (1,000 lbs. = one kilopound or kip):

Live load: 8,000 lbs./ft
Dead load: 39,000 lbs./ft..
Total (q): 47,000 lbs./ft.

$$H = \frac{qL^2}{8d} = \frac{(47,000)(3500)^2}{8(325)} = 221,442 \text{ kips}$$

Length (L): 3,500 ft
Depth (d): 325 ft..

$$V = \frac{qL}{2} = \frac{(47,000)(3500)}{2} = 82,250 \text{ kips}$$

Four parallel cables hold up the roadway, two on each side. For any given point, tension in a cable (T) is given by the formula $T = \sqrt{H^2 + V^2}$. For the cable inside the towers, $T = 236,223$ kips. In the cables leading to the abutments (T_1), on the New Jersey side $T_1 = 260,000$ kips and on the New York side $T_1 = 261,000$ kips. To be safe, Ammann took 261,000 kips for the tension in the entire length of the cable from abutment to abutment.

Steel has a yield point at which it begins to deform and a point at which it will break. Both are measured in pounds per square inch (psi). Ammann chose a cable with a breaking point of 220,000 psi and a yield point of 150,000 psi. An engineer normally designs a cable to withstand one-half of the yield point, a factor known as the allowable stress (f_a). Dividing the tension by the allowable stress gave Ammann the total cross-sectional area (A) of the four cables. He took 82,000 psi as the allowable stress.

$$A = 261,000 \text{ kips}/82 \text{ psi} = 3190 \text{ sq. in. (rounded to 3200 sq. in.)}$$

$$= 800 \text{ sq. in. in each cable}$$

Source: Allston Dana, Aksel Anderson, and George M. Rapp, "George Washington Bridge: Design of Superstructure," *ASCE Transactions* 97 (1933): 105–10.

Sidebar 8.5　**The Deflection Theory**

The deflection theory relates the live load on the cable and the deck of a suspension bridge (P), the vertical distance that the cable and deck "deflect" or descend under this weight (v), and the stiffness of the cable and deck (S). The diagram simplifies the mathematics in the theory in order to explain its basic idea:

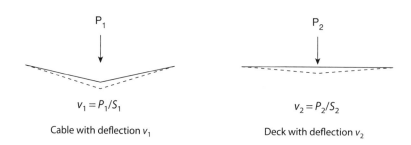

$$v_1 = P_1/S_1$$

Cable with deflection v_1

$$v_2 = P_2/S_2$$

Deck with deflection v_2

A live load (P) consists of the proportion carried by the cable at its midpoint (P_1) and the proportion carried by the deck at its midpoint (P_2). The deflections of the cable (v_1) and of the deck (v_2) occur at the same point and so descend the same distance: thus $v_1 = v_2$. The cable and deck also possess stiffness, which for the cable can be designated S_1 and for the deck S_2. The amount of deflection equals the weight divided by the stiffness: $v_1 = P_1/S_1$ and $v_2 = P_2/S_2$.

In place of $v_1 = v_2$, we can substitute the equation $P_1/S_1 = P_2/S_2$. Multiplying each side by S_1, we can then say that $P_1 = P_2(S_1/S_2)$. Substituting $P_2(S_1/S_2)$ for P_1, the total live load, $P = P_1 + P_2$, can be rewritten as $P = P_2(S_1/S_2) + P_2$, or $P = P_2(1 + S_1/S_2)$. Dividing each side by $(1 + S_1/S_2)$ gives us the formula: $P_2 = P/(1 + S_1/S_2)$.

From this formula, $P_2 = P/(1 + S_1/S_2)$, it can be seen that if the deck stiffness (S_2) is 100 times more than the stiffness of the cable (S_1), then $S_1/S_2 = 1/100$, and $P_2 = P/(1 + 0.01)$. The live load on the deck P_2 amounts to all but about 1 percent of the load. But if the cable stiffness is 100 times more than the deck, then $S_1/S_2 = 100/1$ and therefore $P_2 = P/(1 + 100)$. The cable carries all but about 1 percent of the load.

Ammann and other engineers of the time concluded that if almost all of the stiffness was in the cable, almost none would be needed in the suspension bridge deck. Unfortunately, deflection theory did not take into account the dynamic effects of winds on a deck with very little stiffness. After the Tacoma Narrows Bridge failure of 1940, suspension bridges were designed with stiffer and more aerodynamically stable decks.

suaded Ammann to think that if the cable carried a great dead load, the deck would need almost no stiffness and could be designed to have a very light and slender form (sidebar 8.5).[23]

Ammann's original design included vertical trusswork between the two decks (figure 8.2). The deflection theory encouraged him to eliminate the lower deck and vertical truss altogether in the final 1931 bridge, so that the cables suspended only a single thin roadway for cars, trucks, and pedestrians (figure 8.3). Estimated use of the bridge made the lower deck unnecessary, and the deflection theory provided a structural reason for not building it. The final cost of the George Washington Bridge was $3 million less than the $60 million in bonds sold by the Port Authority to pay for the project.[24] Ground breaking began on September 21, 1927, and the bridge opened for traffic on October 25, 1931.

Suspension bridges in the United States began to be designed to have lighter decks in 1909, but the George Washington Bridge set a new standard in the design of large-scale bridges.[25] Later cable bridges in the 1930s followed its example of suspending very thin decks over very long spans.[26] The decks of the later structures soon ran into trouble. On a suspension bridge, the cables resist only vertical loads. The deck is vulnerable to horizontal and vertical motion caused by winds. Bridge engineers, following Ammann, regarded wind as a straight horizontal force that could be resisted by flat trusswork underneath the deck. But dynamic winds can gust under and over a bridge in ways that lift and depress the roadway deck. A very thin deck can undulate or gallop under severe wind conditions, and horizontal trusses will not provide enough protection.

Bridges built later in the 1930s with thin but narrower decks experienced dangerous problems with winds. Four months after its completion in 1940, the Tacoma Narrows Bridge in Washington failed when moderate winds caused the structure to lift up, undulate, twist, and then collapse (figure 8.4).[27] In the aftermath of Tacoma, structural engineers added vertical stiffness to the decks of suspension bridges that were at risk. The George Washington Bridge was so wide and so heavy that dynamic winds did not make its deck unstable. In 1962 a second lower deck was added to accommodate increased traffic, and this double deck (for which the cables were originally designed) added stiffness to the span. Ammann went on to design a number of other large bridges, including an arch bridge between Bayonne, New Jersey, and Staten Island, New York. His last and

Figure 8.2. Artist's conception of the George Washington Bridge, with masonry. Courtesy of the American Society of Civil Engineers, Washington, DC. *ASCE Transactions*, 97 (1933): 53.

Figure 8.3. Artist's conception of the George Washington Bridge, with a single deck. Courtesy of the American Society of Civil Engineers, Washington, DC. *ASCE Transactions*, 97 (1933): 55.

Figure 8.4. Collapse of the Tacoma Narrows Bridge, November 7, 1940. Courtesy of Special Collections, University of Washington Library, Seattle, WA. No. UW 21422.

greatest structure was a suspension bridge over the Verrazano Narrows between Brooklyn and Staten Island, completed in 1964. For the Verrazano Bridge, Ammann developed a tubular deck that provided greater stiffness without the box-like appearance of a truss. Since then, large suspension bridges have been built with thin decks that are aerodynamically designed to be safe against wind.[28]

History and Aesthetics in Design

Many engineers believe that the engineering of the past has no relevance to the present. Ammann did not take this view. He saw his work as part of a tradition of graceful long-span bridges that began with the English suspension bridges of the early nineteenth century.[29] However, further historical study would have revealed to him the experience of the earlier bridges with dynamic winds and the failure of some as a result. Thomas Telford's Menai Straits Bridge in Wales had to be reinforced with vertical trusses under the deck. John Roebling built a different kind of reinforcement into the Brooklyn Bridge by using cable stays, diagonal wires extending from the towers to the deck. These stays were more elegant than trusses and appeared to many to be ornamental, but their real purpose was to help stabilize the roadway against wind.[30]

In the design of large bridges, civil engineers in the twentieth century tried to make their work more scientific and mathematically sophisticated. Although the George Washington Bridge proved safe, the deflection theory misled designers of narrower long-span bridges in the 1930s and led to the collapse of the Tacoma Narrows Bridge in 1940. Although daring, Ammann's traffic-load estimate was informed by experience, and the flow of traffic across the bridge proved it to be safe.

Nineteenth-century bridges usually had masonry towers or were covered entirely in stone, because tradition demanded a stone appearance in public works. The steel towers on the George Washington Bridge were designed to be covered with concrete and then faced with stone. The total project cost of $57 million (the bridge, approaches, and other expenses) did not include this masonry work. To keep down the cost, the Port Authority decided to leave the steel towers uncovered, and the public embraced the skeletal towers as a striking visual feature of the bridge. The George Washington Bridge thereby symbolizes an ambiguity. Its dense steel framework represented the structure required to support both the cables and the structurally unnecessary concrete and stone. Ammann's later bridges were free of such excess steel.

Viewed from a distance, however, the George Washington Bridge expresses an unambiguous striving for elegance. In the formula for the horizontal force, $qL^2/8d$, the span length L and the depth of the cable sag d were matters of choice. The bridge would have performed just as efficiently and its cost would have been about the same had Am-

Figure 8.5. The George Washington Bridge today. Courtesy of J. Wayman Williams.

mann designed it to have taller towers (higher d). Such a design would have added to the steel needed in the towers but reduced the amount of steel needed in the cables.[31] But a bridge with taller towers would have been ungainly. By choosing shorter towers and a flatter cable sag, Ammann gave the bridge a more graceful appearance (figure 8.5).

In the design of large bridges, aesthetics are not an add-on to the design or an added cost; they are intrinsic to the conception of the structure. An understanding that efficiency, economy, and elegance are all required by good design was a vital part of the vision that Ammann brought with him from Switzerland.[32] His 1931 Bayonne Bridge beat out the George Washington Bridge for the aesthetics award that year of the American Institute of Steel Construction. But both bridges displayed Ammann's ideal of aesthetic excellence as an integral part of an efficient and economical design. Ammann would go on to design other major New York crossings, including the Bronx-Whitestone, the Throgs Neck, and the Verrazano Narrows Bridges. The Bronx-Whitestone Bridge, completed in 1939, was built in conjunction with the 1939 New York World's Fair and is often regarded as Ammann's most elegant design.

In his successful effort to build what was at the time the world's longest-spanning bridge, Othmar Ammann operated as an entrepreneur of structure in steel. Another new building material, reinforced concrete, also opened new entrepreneurial possibilities for structure in the first four decades of the twentieth century. Two engineers, John Eastwood and Anton Tedesko, explored the potential of reinforced concrete to achieve striking new forms.

CHAPTER NINE

Eastwood, Tedesko, and Reinforced Concrete

Along with the George Washington Bridge in New York and the Golden Gate Bridge in San Francisco, the most famous American public work of the 1930s was Hoover Dam, on the border between Nevada and Arizona.[1] Made of a new material, modern concrete, the dam was the largest concrete structure in the world when it was finished in 1936. But Hoover Dam was not an example of efficient design. Built when materials and labor were cheap, the dam used far more concrete than it needed to be safe, and its massive look did not take advantage of concrete's potential for elegant thinness.

During the first four decades of the twentieth century, the works of two civil engineers, John Eastwood and Anton Tedesko, realized the potential for elegance and economy in reinforced-concrete design. Eastwood and Tedesko stressed the idea of *form* over *mass*, the idea that the amount of mass needed in a structure could be reduced safely by the use of an efficient form. The civil engineering profession was at first suspicious of their approach. The kinds of structures involved, thin dams and roof shells, are today rarely built, although most of those already in use have served well with proper maintenance. Ultimately, the profession recognized the validity of what Eastwood and Tedesko accomplished. The idea of form over mass, and the related idea that efficiency, economy, and elegance must all be present in a good structural design, still speak powerfully to the needs of public works and private buildings today.

In the period from 1876 to 1939, the new networks of electric power and telephony, the processes of oil refining, and the making of machines such as the automobile came under the control of a small number of large business organizations. In the engineering of large-scale bridges and buildings, though, large private firms did not dominate. Bridge designers usually worked for public authorities as consultants or as staff engineers, while designers of buildings and other structures usually worked as consultants.

Eastwood and Tedesko worked as consultants in private practice. Unlike the great names associated with machines and networks, innovators of new structures were not well known. But in their works, Eastwood and Tedesko exemplified the qualities that characterize outstanding technology.

Mass versus Form in Reinforced Concrete

Roman builders made concrete by mixing a natural cementing agent with water, sand, and crushed stones. During the nineteenth century, a new and better cement came into use, an industrially produced agent called Portland Cement, that made a stronger concrete. Embedding steel rods in the new concrete enabled it to resist higher tension and created a new building material, reinforced concrete. The materials used to make reinforced concrete are inexpensive and readily available in most parts of the world, so that building in the new material is now a major part of the construction industry.[2]

Concrete is custom-made at the building site for most projects and is therefore a less well-controlled material than manufactured steel. But once engineers could assure themselves that they could control the quality of the material during construction and see it perform safely in use, they began to design in reinforced concrete. At first, most engineers saw it only as a cheaper substitute for stone and used it to build heavy, stone-like structures for dams, bridges, and buildings. But a few engineers saw the potential for reinforced concrete to carry loads through new forms that were not practical in stone.

The first engineer to realize fully the possibilities of reinforced concrete was the Swiss structural engineer Robert Maillart (1872–1940). In 1930 he wrote: "It is usual to believe that massive structures are necessarily strong. Mighty pillars and thick arches arouse confidence in the mind of the observer, while light membered structures cause more anxiety than delight." But, he added, "in antique remains firm slender columns are often found next to collapsed mass masonry" and "there is no doubt that slender structures are just as beautiful to the eyes of the layman and even more beautiful than massive forms."[3] In his bridges, Maillart progressively dispensed with mass in favor of lighter and more slender forms that controlled the way in which the structure carried load. In 1947 the New York Museum of Modern Art held an exhibition of his work, the first on the theme of structural engineering as art.[4] The highlight of this exhibition was

Maillart's Salginatobel Bridge, the first reinforced-concrete bridge to become an International Historic Civil Engineering Landmark.[5]

John S. Eastwood and Anton Tedesko worked to achieve efficient, economical, and elegant structures of reinforced concrete in the United States. Eastwood devoted the latter part of his life to dams; Tedesko spent the first part of his career designing thin-shell roofs. Both stood for the idea of form over mass.

Eastwood and the Multiple Arch Dam

John S. Eastwood (1857–1924) was born on a farm near Minneapolis, attended the University of Minnesota without completing a degree, and then worked as a surveyor and construction engineer (figure 9.1). He settled in Fresno, California, in 1883. In 1895 he helped build a hydroelectric system for the San Joaquin Electric Company. During a drought in 1899, the company failed because it did not have a reservoir large enough to store water to power the electrical network. Eastwood realized that dam building could play a crucial role in the often arid American West.[6]

Most dams in the United States at that time were *earth* dams, consisting of an earth wall, sloped on each side, to block the flow of a stream (sidebar 9.1). Larger earth dams had a wall of rock, clay, or concrete in their core to prevent seepage. A second type of dam, called *gravity* dams by engineers, consisted of an exposed wall of rock or concrete, sloped outward mainly on the downstream side. A variant of the concrete gravity dam was the Ambursen "flat slab" dam, which saved concrete by making a relatively thin slab sloped on the upstream side with widely spaced perpendicular buttresses on the downstream side. A third major kind of dam was the *arch* dam, in which the structure's shape provided most of its strength. An arch dam worked like an arch bridge laid on its side, with the outside of the arch resisting the water. An arch dam needed much less material than a gravity dam to hold back the same amount of water.[7]

At the turn of the century, wealthy investors saw the potential for California to grow if water and electric power could be supplied from mountains in the interior to cities along the coast. Dams were needed to store water for drinking and irrigation and for powering electric generators. In 1901 Henry Huntington, whose uncle was a founder of the Central Pacific Railroad, began a system of power stations and transmission lines

Figure 9.1. John Eastwood. Courtesy of Michigan State University Archives and Historical Collections, East Lansing, MI.

Sidebar 9.1   **Three Types of Dam Design**

Earth Dams

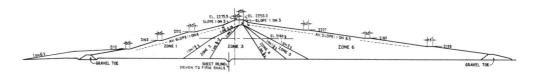

Fort Peck Dam, Montana (elevation)

The larger earth dams usually have embankments of earth and rubble and a core of more impervious material. The dams have a gradual slope and great mass.

Gravity Dams

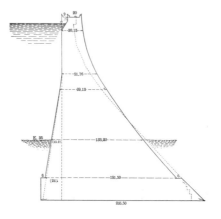

So-called gravity dams are made of masonry, rock, or concrete. These dams resist water pressure by their great weight and harder material. They are usually smaller than earth dams but also rely on their mass to hold back water.

New Croton Dam, New York (elevation)

Arch Dams

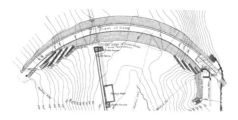

Arch dams rely on their form to provide strength and require less mass than gravity dams. An arch dam can have one arch or many.

Sweetwater Dam, California (plan view)

Sources: U.S. War Department, Corps of Engineers, U.S. Army, *Report on the Slide of a Portion of the Upstream Face of the Fort Peck Dam* (Washington, DC: U.S. Government Printing Office, 1939), fig. 6 (after p. 14), for Fort Peck Dam; *ASCE Transactions* 58 (1907): 425 for New Croton Dam; *ASCE Transactions* 19 (1888): pl. 28 (after p. 218) for Sweetwater Dam.

to bring electricity to Los Angeles.[8] In 1902 Eastwood proposed three dams to impound a reservoir in Big Creek, a tributary of the San Joaquin River, as part of this system.

Eastwood envisioned a new kind of dam, one that would employ multiple arches of reinforced concrete. For each dam at Big Creek, he designed several arched walls, each forty feet wide, with buttresses on the downstream side for support. The three dams would require 73,000 cubic yards of concrete, more than the 64,000 needed by core walls for earth dams. But the design did away with 1 million cubic yards of earth. A traditional concrete gravity design would have required about 300,000 cubic yards of concrete. Eastwood calculated that an arched concrete wall could hold back as much water as a far heavier flat wall of concrete. But Huntington and his engineers commissioned three concrete gravity dams at greater expense instead.[9]

While waiting for approval at Big Creek, Eastwood had the chance to build two arch dams in California. In 1908, to help a lumber company hold and transport logs, Eastwood built the Hume Lake Dam in the Fresno region, the first ever built to use multiple arches of reinforced concrete (sidebar 9.2 and figure 9.2). The dam cost $46,000, far less than the $70,000 estimated for a gravity design that would have used rockfill. The Hume Lake Dam was 667 feet long, 61 feet high at its highest point, and contained twelve arched walls, each 50 feet wide. The arches were sloped on the upstream side and supported on the downstream side by buttresses.[10]

In 1910 the Bear Valley Mutual Water Company commissioned Eastwood to build a dam to store water for irrigation in the San Bernardino region of southern California. His design showed the aesthetic as well as technical possibilities of using form rather than mass in reinforced concrete. Built in 1910–11, the Big Bear Dam was 363 feet long, 80 feet high, and had ten arches each spanning 32 feet. The buttresses were 18 inches thick at the top and widened to more than 5 feet at the foundations. Because of the greater height at Big Bear (compared with that at Hume Lake), Eastwood designed every buttress to be braced laterally by bridge-like arches and frames between the buttress walls. From a distance, this gave an elegant lattice-like appearance to the downstream side of the dam (figure 9.3).[11]

Eastwood soon found a chance to build a much larger dam using multiple arches. The Great Western Power Company wanted a hydroelectric power dam at Big Meadows, fifty miles north of Sacramento in northern California. Eastwood sent a pro-

Figure 9.2. Hume Lake Dam, near Fresno, California. Courtesy of the Prints and Photographs Division, Library of Congress, Washington, DC. HAER-CA-16-1, Historic American Engineering Record.

posal in early 1911 to H. H. Sinclair, the company vice president, for a multiple-arch design 720 feet long and 150 feet high with twenty-two arches, each 30 feet wide. Sinclair sent Eastwood's proposal to James Schuyler, a leading authority in the West on dam design. Schuyler approved Eastwood's design and noted that it would cost only about $600,000 compared with about $1.2 million for a gravity dam. Alfred Noble, an expert in New York, gave a similarly favorable review.[12]

Sidebar 9.2 **Hume Lake Dam**

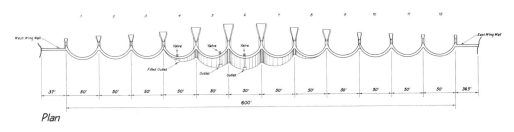

Plan

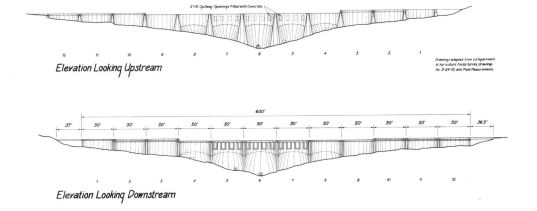

Elevation Looking Upstream

Elevation Looking Downstream

John Eastwood's 1908 Hume Lake Dam was the world's first reinforced-concrete, multiple-arch dam. Each arch worked like an arch bridge laid on its side, holding back the water of the Ten Mile Creek near Fresno, California.

The Hume Lake Dam is 677 feet long, 61 feet high at its highest point, and has twelve arched walls that are each 50 feet wide. Buttresses support the walls on the downstream side. The arched walls slope outward as they descend on the upstream side, giving added resistance to water pressure.

With 2,200 cubic yards of material, the Hume Lake Dam had a total storage capacity of 1,530 acre-feet of water. The 1905 New Croton Dam, a gravity dam north of New York City, used 1,450,000 cubic yards of masonry to achieve a total storage capacity of 86,330 acre-feet of water. The Hume Lake Dam had a ratio of storage to mass of 0.69 while the New Croton Dam had a ratio of 0.05.

Sources: Donald C. Jackson, *Building the Ultimate Dam: John S. Eastwood and the Control of Water in the West* (Lawrence: University Press of Kansas, 1995), pp. 86–89; Historic American Engineering Record, Library of Congress, Washington, DC.

Figure 9.3. The Big Bear
Dam. Courtesy of Professor
Donald C. Jackson,
Lafayette College.

Then disaster struck. Sinclair had to resign temporarily for health reasons, and the company president died. Although Sinclair returned, the new president called in John R. Freeman (1855–1933), a leading civil engineer and dam designer, for advice. Freeman believed that massive dams were the only ones that were safe. He also insultingly characterized Eastwood as a designer who "has become so impressed with the beauties of his multiple arch that I presume he will build his house shingled with semicircular tiles and ultimately have his hair trimmed in scallops."[13] At Freeman's urging, work on Eastwood's dam stopped in October 1912 after $188,500 had been spent on the concrete and $146,000 on excavation. In March 1913 the company abandoned the design and built a massive earth dam in its place.[14]

Eastwood rebounded, though, when he became a consultant to an open competition for the Mountain Dell Dam near Salt Lake City, Utah. The State of Utah asked Eastwood to submit a design for a multiple-arch dam and invited other designers to submit plans for a gravity dam and an Ambursen dam. Contractors then submitted bids estimating what they would charge to build each design in two stages, a "base" dam 110 feet high and a completed dam 150 feet high. Eastwood's multiple-arch design won with bids of $75,300 for the base and $139,000 for the completed dam. The lowest bids for the gravity dam were $88,700 and $226,000 respectively, and for the Ambursen dam $117,000 and $217,000. The Mountain Dell competition proved that Eastwood's ideas could triumph when given the chance in open competition with rival methods.[15]

Eastwood was able to build ten more dams before his death in 1924. The largest of these was the Lake Hodges Dam, begun in 1917 on the San Dieguito River north of San Diego.[16] As the dam neared completion in March 1918, a flood sent water through the opening of an unbuilt arch. The heavily loaded structure suffered no damage, dispelling a worry that multiple-arch dams would fail if one of the arches failed or had not yet been built. In his Littlerock Dam in the Mojave Desert, finished in 1924, Eastwood carried his aesthetic vision further by proposing a radial design, in which the arches were arranged along a curved rather than straight line. Objections from state engineers led to the construction of a straight dam, still an elegant structure of Eastwood's design.[17] At Webber Creek, about fifty miles east of Sacramento, he designed a very thin 115-foot high dam with a central arch 140 feet wide and two arches 105 and 115 feet in span. The dam was built only up to 90 feet in height.[18] In his later dams, Eastwood achieved a level of efficiency not equaled since in American dam design.

On March 12, 1928, the St. Francis Dam north of Los Angeles collapsed cata-strophically, killing more than 400 people.[19] Although this was a gravity dam built by the City of Los Angeles, its failure prompted the California legislature to pass a law placing all dams in the state under the supervising authority of the state engineer, who asked the engineer Walter Huber, a follower of John Freeman, to head a "multi-ple arch dam advising committee." Huber's committee arbitrarily decided that East-wood's Lake Hodges Dam was unsafe, and in 1936 engineers added concrete rein-forcement into the openings between every other pair of buttresses. The reinforcement resembled the cross-bracing in the towers of the new Golden Gate Bridge in San Francisco, but its use in the dam was not the result of any evidence pointing to a need for it.[20]

During the 1930s California and other U.S. states built very narrow long-span sus-pension bridges following the deflection theory, even though the history of smaller nineteenth-century bridges of similar dimensions showed the susceptibility of such bridges to failure in wind.[21] At the same time, new dams were built to be massive. A 1920 survey of one hundred dam failures showed that no thin arch dam had ever failed structurally; nearly all failures were of massive masonry or earth dams.[22] Yet the engineering profession believed arch dams to be unsafe. The Tacoma Narrows col-lapse in 1940 showed the error in 1930s suspension bridge design, and time has vindi-cated the safety of Eastwood's multiple-arch dams. Although a few have been rehabili-tated, none has ever failed.

Anton Tedesko Comes to America

The idea of form over mass also developed in Europe in the pioneering design work of Dyckerhoff & Widmann, an engineering and construction firm in Wiesbaden, Ger-many. Working in reinforced concrete, the firm experimented with new ways to cover large spaces in the 1920s. The firm built domes and cylindrical "barrel" shells to serve as large roofs of extraordinary thinness.[23] The possibilities of thin-shell design using re-inforced concrete fascinated an Austrian civil engineering graduate, Anton Tedesko (1903–94), who joined the firm in 1930 after spending two years working in the United States (figure 9.4).[24]

In 1932 the German firm sent Tedesko to America to introduce thin-shell concrete roof construction there. The Chicago office of the Roberts and Schaefer Company agreed to give him office space. Over the next two years, Tedesko met with engineers, architects, builders, and owners in an effort to promote the new German thin-shells, under the name Z-D (Zeiss-Dywidag) Shell Roofs.[25] But he faced the worst possible conditions in which to argue for new construction: the deep economic depression of the early 1930s and a conservative civil engineering profession. In mid-1932 Tedesko traveled to New Orleans to design a large arena in collaboration with a local architect and engineer. After long hours over a drafting board in the sticky delta summer, Tedesko produced a concrete shell design only to see it rejected in favor of a conventional steel truss roof.[26] Like Eastwood at Big Creek, Tedesko's first effort was beaten by a standard design: massive dams and steel trusses were typical American solutions that engineers preferred in the 1920s and 1930s to exotic-looking thin concrete arches and shells.

Tedesko met with success, however, at the Century of Progress World's Fair in Chicago of 1933–34. He won a contract to design and build a thin-shell concrete roof for the Brook Hill Dairy Exhibit. On American farms, dairy cows were housed in wooden barns. Tedesko convinced Brook Hill's architect and consulting engineers to accept the unorthodox idea of a pavilion in reinforced concrete. The engineers advertised the resulting structure as giving cows "comfort and safety greater than that ever before enjoyed by any cows anywhere" in a building that "cannot burn, rust, rot, or blow away."[27] The dairy structure had a barrel-shaped roof and tests of its load-carrying capacity impressed engineers from the the University of Illinois and the Portland Cement Association (the trade group of cement manufacturers in the United States), as well as representatives of the Roberts & Schaefer Company.[28] The roof for cows did not cause the rush to shells that Tedesko had hoped to see. But Tedesko found new work and in 1934 he became an employee of Roberts & Schaefer, although the company that year reduced salaries by 45 percent.

The Hershey Arena

Tedesko's breakthrough as a designer came later in 1934, when he designed the dome of the Hayden Planetarium in New York City (between 77th and 81st streets along

Figure 9.4. Anton Tedesko in 1936. Courtesy of Princeton Maillart Archive, Princeton University, Princeton, NJ. Tedesko Papers.

Central Park West in Manhattan). This was the first reinforced-concrete thin-shell dome to be built in the United States; it was eighty feet and six inches in diameter and only three inches in thickness (figure 9.5).[29] Tedesko made a convincing case for the economy and safety of the 1920s domes in Germany. He also benefited from worries that New York was falling behind other U.S. cities. Planetaria had been built in Philadelphia and Chicago, and one was under construction in Los Angeles. New York had no planetarium and was eager to have this new means of projecting the images of celestial bodies in motion.

Tedesko's greatest opportunity came in 1935, though, when the Philadelphia representative of Roberts & Schaefer, James Gibson, introduced him to Milton Hershey (1857–1945), who owned the largest chocolate-making plant in the world. In 1903 Hershey had built the town of Hershey, Pennsylvania, around his factory, where he had

Figure 9.5. The Hayden Planetarium. Courtesy of Princeton Maillart Archive, Princeton University, Princeton, NJ. Tedesko Papers.

also founded a school for orphans. The Depression hurt the chocolate industry and in 1935 Hershey decided to build a sports arena for ice hockey as a way to keep his workers employed.[30] On January 21, 1936, Tedesko submitted a proposal to design and build a thin concrete shell to enclose the arena. The Hershey Company accepted with the provision that he employ chocolate workers to do the construction. Tedesko agreed.[31] Work started immediately with design staff hired in Chicago, and on March 11 chocolate workers began to dig the foundations of the structure. The Hershey Arena would cover an area 340 feet long and 232 feet wide. For the oblong arena, Tedesko designed a thin reinforced-concrete, barrel-shell roof, three and one-half inches thick, supported across its width by eight arches. He did not design cross-ribs between the arches in the belief that they were unnecessary.[32]

A roof posed a challenge different from that of a bridge or dam. On a bridge, live load from traffic is significant, and a dam must resist the live load of water on its upstream face. On a long-span concrete roof, live load (mainly rain and snow) is a small fraction of the dead load of the structure itself. Tedesko realized that the supporting arches did not need to be of uniform depth. Half of the roof's weight would fall on the abutments, and he designed the arches to be twelve feet deep at these points, where they merged into the side walls of the arena. The arches tapered to a depth of five feet at the crown, giving a graceful appearance as well as the means for stiffening the thin-shell roof. Although it was only three and one-half inches thick, the shell roof was strong enough to carry its own weight and the weight of any live load, even without the supporting arches. (sidebars 9.3 and 9.4).[33]

Tedesko designed the arches to be able to support the entire roof load, including the thin shell and all of the live load. He also made calculations to show that the thin shell could carry its own weight and live load without help from the arches, except near their lowest edges. It was thus a conservative design, but one which still showed the possibilities for supporting wide spans with relatively small amounts of material.

In addition to an efficient design of the structure, Tedesko had to develop a construction process that could result in reasonable economy. He designed a reusable wooden scaffolding that could support one unit of two arches with connecting thin shells. After the concrete had been cast and had hardened, the scaffolding was lowered and moved to an adjacent unit. Removing the wood required carefully lowering the supports beneath the arches in a pre-designed sequence that Tedesko controlled from a central station (figure 9.4). The concrete had to be stiff so that it could be cast on steeply sloped forms without flowing down (figure 9.6). The final result showed the thinness of shells and arches (figure 9.7), which would be partly concealed when the side walls of the arena were finished (figure 9.8).

• •

Sidebar 9.3 **The Hershey Arena: Horizontal Forces**

Anton Tedesko designed the 1936 Hershey Arena in Hershey, Pennsylvania, as a thin reinforced-concrete shell stiffened by arches. Normal reinforced concrete will crush under a compressive stress of about 3,000 pounds per square inch (psi), so he designed the stresses in each arch and in the roof to fall far below this amount.

He estimated the total load on each arch and abutment to be 867,000 pounds or 867 kilopounds or kips. Because the arch is much heavier near the abutments, more than half of the weight or 464 kips would fall directly over the abutments and would not need to be supported by the arch span itself. Each arch therefore needed to carry roughly 403 kips. With a span length (L) of 220 feet, the weight per foot of span length (q) was 1.83 kips ($1.83 \times 220 = 403$). A rise (d) of 81 feet at the crown gave a horizontal force of 136 kips in each arch at the crown:

$$H = \frac{qL^2}{8d} = \frac{(1.83)(220)^2}{8(81)} = 136 \text{ kips}$$

The force in the arch increased as the arch descended to the abutments, but Tedesko made the arch deeper there. The shell roof extended outward (at right angles to the arches), and Tedesko assumed that the shell's dead load also fell onto the arch. This shell load came to 2 kips per foot and added 149 kips to the horizontal force:

$$H = \frac{qL^2}{8d} = \frac{(2.0)(220)^2}{8(81)} = 149 \text{ kips}$$

Finally, an estimated 0.98 pounds per foot of snowfall added 73 kips of live load. Combining these three weights ($136 + 149 + 73$ kips) gave a total H of 359 kips.

Sources: Edmond P. Saliklis and David P. Billington, "Hershey Arena: Anton Tedesko's Pioneering Form," *ASCE Journal of Structural Engineering* 129, no. 3 (March 2003): 278–85. We simplify here by assuming that the arch and the shell rested on three hinges (two abutments and a crown or midpoint), although the actual arches and roof had only two hinge points (the two abutments). We also assume that the weight across the arch and roof shell was uniformly distributed.

• •

The Hershey Arena was the largest thin-shell concrete roof in America, covering 78,880 square feet. It demonstrated that such a huge area could be covered by a concrete surface of extraordinary thinness, primarily by taking advantage of the curved shell form. For Tedesko this was "the most satisfying challenge of the 1930s. . . . The engineering and construction decisions were mine. No codes existed that would apply

- -

Sidebar 9.4   **The Hershey Arena: Stresses**

With a horizontal force of 359 kips, Tedesko next measured A, the cross-sectional area of the arch crown. At this location, multiplying the depth (5 feet) by the width (1.83 feet) gave an area of 9.15 square feet. Converting 9.15 square feet into square inches (multiplying by 144) gave a cross-sectional area of 1,317.6 square inches for a stress of 272 psi. The stress (f) in the arch of 272 psi was far below the crushing stress of 3,000 psi for the reinforced concrete:

$$f = H/A = 359,000/1317.6 = 272 \text{ psi}$$

For added safety, though, Tedesko designed the entire thin-shell roof to resist crushing under its own weight and the weight of snow, even without the arches to support it. Another formula gives the stress in a shell:

$$f = pR/h$$

in which p is the combined dead and live load in pounds per square foot (psf), R is the radius of the shell in feet, and h is the thickness of the shell in feet.

For the Hershey Arena, the shell load was 73 psf, the radius was 125 feet, and the thickness was 3.5 inches or 0.291 feet. Dividing the result of 31,357 psf by 144 gave a shell stress of 217 psi, well below the concrete's crushing stress of 3,000 psi:

$$f = \frac{qR}{h} = \frac{(73)(125)}{0.291} = 31,357 \text{ psf}/144 = 217 \text{ psi}$$

A thin-shell roof can buckle under compression, though, at a lower stress than the crushing stress of concrete. For his thin-shell roof, Tedesko estimated the buckling stress to be 990 psi. An engineer would normally design a shell to be less than one-third of the buckling stress, or 330 psi here. The shell stress of 217 psi was well below this point.

Sources: Saliklis and Billington, "Hershey Arena"; David P. Billington, "Anton Tedesko," *Journal of the Structural Division, Proceedings of the ASCE* 109, no. ST11 (1982).

- -

Figure 9.6. (Right) Hershey chocolate workers casting concrete. Courtesy of Princeton Maillart Archive, Princeton University, Princeton, NJ. Tedesko Papers.

Figure 9.7. Hershey Arena arch, half cross section. Courtesy of Princeton Maillart Archive, Princeton University, Princeton, NJ. Tedesko Papers.

Figure 9.8. Completed Hershey Arena in use. Courtesy of Princeton Maillart Archive, Princeton University, Princeton, NJ. Tedesko Papers.

to this work. No rules had to be followed. I shaped and calculated the structure according to my best judgment, influenced by what I had learned in Wiesbaden." The Hershey Arena gave Tedesko a unique chance to develop independent judgment and self-confidence. "In Europe, there would not have been only a single person in charge of such a project, and certainly not someone 32 years old."[34]

Tedesko faced a crisis on July 3, 1936, during the concrete casting for the first roof section of the Hershey Arena. As he would describe years later, "we started understaffed, had trouble with concrete aggregate, with breakage of form supports. My instructions that tarpaulins be available to cover fresh concrete were ignored. Concrete started to set prematurely under the hot sun; I added water, and the mix became too wet." Then: "At 2:30 a.m. hell broke loose with an electric storm and torrential rain. The men dropped their tools and fled to covered areas." What happened next was a concrete engineer's worst nightmare. "Rivers came down the arched roof and washed down the concrete which had not yet hardened. Concrete mud slides came to rest on the flat side roofs, not strong enough to support them, and I assumed that the increasing load of the ponding water (perhaps 90 tons) would lead to collapse." Tedesko worked frantically, first trying to siphon off the water but "finally breaking holes through the young concrete of the flat side roofs, through which gravel, concrete mud, and rain water could drop, relieving the dangerous load. . . . I was dirty and wet, but I felt good, like a conqueror. I could also see the humor of the situation."[35]

The Hershey people saw no humor, though, and were sure that the structure would have to be torn down and a standard steel truss roof built instead. But Tedesko confidently promised to make all right. After seventy-two hours of round-the-clock work to clean up the mess, Tedesko was ready to begin the concrete placement again. This time all went well and soon word spread around the county that something unusual was arising in the green rolling hills of the Susquehanna Valley. Visitors came, especially prominent engineers and builders, some of whom would become close friends of Tedesko and help him get future projects. One was Lieutenant-Commander Ben Moreell, who later became a full admiral and an aide to President Roosevelt during the Second World War. Moreell and other officers saw Tedesko's long-span roof as an ideal form for military aircraft hangars. Much of the housing for U.S. warplanes during World War II thus owed their design to Tedesko's chocolate arena.[36]

On one of his numerous train trips between Harrisburg and Chicago, Tedesko noticed everyone looking at him as he sat on the aisle in the dining car. Suddenly he realized that the woman across the aisle was the First Lady, Eleanor Roosevelt.[37] Such were the rewards of his hectic schedule during the last half of 1936.

Later Prewar Shells

Tedesko still kept in touch with his colleagues in Germany, but a trip to Europe in late 1937 persuaded him that Nazi Germany was no place to live. He became a U.S. citizen in 1938 and married an American, Sally Murray, whom he had met in Chicago three years earlier.[38] Americans began to accept thin-shell roofs after the Hershey Arena, and in 1937–38 he began a new series of roof projects as the designer. Three of these characterize Tedesko's mastery of diverse forms in American conditions. The first was an arched Hershey-like roof shell for the Philadelphia Skating Club and Humane Society in Ardmore, Pennsylvania. Charles Schwertner, who had helped Tedesko on the site at Hershey, became the building contractor for Ardmore and insured a well-built shell with arches spanning 116 feet in a building 235 feet long. It was, according to Tedesko, "the best-looking shell structure of the 1930's . . . attained through joint efforts with architect Nelson Edwards and contractor Schwertner."[39]

Tedesko worked under another Austrian at Roberts & Schaefer, John Kalinka, who in 1937 hired a third Austrian engineer, Eric Molke, to fill in for Tedesko while the latter was in Europe. Molke worked with Tedesko on two dome roofs covering trickling filters for a water treatment facility in Hibbing, Minnesota.[40] These concrete domes were an unusual choice. Trickling filters are usually open but had to be enclosed at Hibbing because of the dangers of icing and heavy snows common in northern Minnesota. The designers made an elliptical (flattened) dome with a ground-plan diameter of 150 feet and a height at midspan of 32 feet. A flat dome shape had the structural advantage of being vertical at the base and thus not requiring a base-reinforcing ring, essential with a spherical dome that would push outward as well as down. The design avoided the more complex mathematical analysis needed for the dome-ring connection.

Barrel roofs are more common than domes because the former can cover rectangular buildings. The largest of the early barrel shells by Tedesko was a roof at Natchez, Mississippi, for an Armstrong tire factory covering 121,600 square feet. Comparative studies of more standard structural types showed that the concrete roof was competitive and reduced the danger of fire. Tedesko had known the local architect through his Chicago fiancée. The main structure was made of cylindrical shells forty feet wide and fifty feet in span. Tedesko calculated the maximum compressive stresses in the shells to be less than 10 percent of the compression capacity of the concrete. Typical for most shells, these results illustrated that thinness did not compromise safety.[41]

By 1939, with these and a number of other contracts in various stages of development, Tedesko had succeeded in proving the value of a new type of roof for large buildings in the United States, using thin reinforced concrete instead of just steel. Tedesko would only reach international fame years later, most notably for his work as structural designer of the Vertical Assembly Building and other structures for the U.S. space program at Cape Canaveral in Florida. Unlike the better-known German rocket engineers who came to the United States after World War II, Tedesko brought his knowledge of German engineering to America before the war. Like Maillart in Switzerland, Tedesko stood for an ideal that did not dominate the design of public structures built in either Germany or the United States in the 1930s: the ideal of form and lightness rather than mass.

CHAPTER TEN

Streamlining: Chrysler and Douglas

N ew form and lightness came not only to structures but also to machines. As automobiles and aircraft increased in speed, they encountered more-severe resistance from the air, or drag. In the 1930s engineers turned their attention to aerodynamic efficiency and began to streamline the shapes of new cars and airplanes. Two machines of the 1930s pioneered streamlined form in the United States: the Chrysler Airflow car and the Douglas DC-3 airplane. The 1934 Airflow proved a commercial failure, in part because its efficient form did not result in an aesthetically pleasing look. It was only half the metaphor of recovery needed by America in the Depression: the car reduced drag but gave no lift. The DC-3 of 1936 was not the first airplane to be streamlined but it was the first passenger plane to be economical as well as aerodynamically efficient. Its powerful lift and reduced drag made the DC-3 the most successful airplane of the decade.

Aerodynamic Form

Streamlining grew out of a deeper understanding of how drag affects a body in wind. Airflows, made visible by the addition of smoke, go over a body in a wind tunnel in the form of streamlines (sidebar 10.1). *Form drag* occurs from the tendency of airflows to separate from the rear surfaces of automobile bodies and airplane wings as they go over them, instead of following the contours of these surfaces closely. Regions of lower air pressure form behind the rear surfaces as a result, increasing the relative pressure of the air in front. Rounding the frontal surfaces and tapering the back ones on cars and on airplane wings and bodies reduce this difference in pressure. Tapered bodies are said to be more streamlined.[1]

Sidebar 10.1 **Streamlining**

Form drag

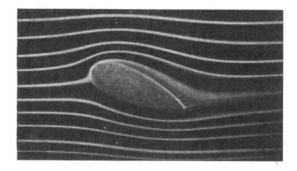

Streamlining to reduce form drag

Adding smoke to airflows in a wind tunnel will make the airflows visible as streamlines. Over an untapered body, airflows will tend to separate, creating a region of low pressure in back of the body and increasing the air pressure in front. The resulting imbalance in pressure is called *form drag*.

Streamlines will follow more closely the contours of a tapered body, reducing the imbalance in pressure. If the body is moving forward, it will meet with less resistance.

The surface of the body will still offer a resistance called *friction drag*. Form drag is the result of air pressure that is perpendicular to a surface. Friction drag is the result of air pressure that is parallel to it and is reduced by smoothing the surface.

Sources: R. C. Binder, *Fluid Mechanics* (New York: Prentice-Hall, 1943), pp. 132, 148 (images); John D. Anderson Jr., *A History of Aerodynamics and Its Impact on Flying Machines* (Cambridge: Cambridge University Press, 1997), pp. 320–28.

During the 1920s and 1930s, the spread of paved roads and more powerful engines enabled cars to go faster. From the Model T's top speed of about thirty-seven miles per hour, cars now reached speeds of fifty and sixty miles per hour and more. Because drag increased by the square of the velocity, automotive engineers realized that cars needed to be designed with greater aerodynamic efficiency in mind. The boxy cars of the 1910s and 1920s began to give way to more streamlined ones in the 1930s. The Chrysler Airflow pioneered this change in the United States.

Walter P. Chrysler

Walter P. Chrysler (1875–1940) established the third great U.S. automobile company in the late 1920s (figure 10.1). Born in Wamego, Kansas, he graduated from high school in 1893 and served an apprenticeship at the Union Pacific Railroad machine shop in Ellis. He then held a series of machinist jobs that moved him across the country to Pittsburgh in 1910, where he became plant manager for the American Locomotive Company. His direction in life began to change, though, when he saw a Locomobile car at the 1908 Chicago Automobile Show. Chrysler went into debt to buy the $5,000 car and learned how it worked. His reputation as a manager who understood machines led General Motors to invite him to manage production in its Buick subsidiary in 1912. Despite a salary of $6,000, half as much as his Pittsburgh employer offered to keep him, Chrysler accepted, and in 1916 he became president of Buick.[2] In 1919 Walter Chrysler took charge of all auto production for General Motors, but he was unable to work under the erratic leadership of Will Durant. In 1920 Chrysler resigned, sold his GM stock, and withdrew from the auto-making business. He was not long to remain on the sidelines, though, and he soon accepted an offer to manage the auto firm of Willys-Overland. In 1921 he bought Maxwell Motors, which became Chrysler Motors in 1925.[3]

The automobile industry was maturing in the 1920s and engineers with more formal training began to play a greater role in automotive design. In 1916 three outstanding university-trained mechanical engineers, Fred Zeder, Owen Skelton, and Carl Breer rescued a new line of cars for the Studebaker Corporation in Detroit by solving performance problems in the cars.[4] In 1920 Chrysler hired Zeder, Skelton, and Breer to design cars for Willys-Overland, and he brought them to work for Maxwell Motors

Figure 10.1. Walter P. Chrysler. Courtesy of Brown Brothers, Sterling, PA.

in 1923, where they designed a new car, the Chrysler Six. Introduced in 1924, the automobile was the first medium-priced car to have four-wheel hydraulic brakes, which made braking safer and easier. The car also had a six-cylinder engine that delivered more power. Chrysler sales of the Six and other cars reached 192,000 in 1927, making the company the fourth largest auto producer in the country.[5]

Unexpected opportunities then enabled Chrysler to become one of the top three auto makers. Henry Ford's last major partners, John and Horace Dodge, sold their shares of Ford Motor Company stock back to Ford in 1919 and built an auto assembly plant of their own in Detroit. But the Dodge brothers died that year, and their widows sold the business in 1925 to an investment banking house in New York. Chrysler offered to buy the Dodge plant for $170 million worth of new Chrysler stock and the bankers accepted. The Dodge plant greatly increased Chrysler's manufacturing capacity. In 1928 the company began mass producing a low-priced car, the Plymouth, and a medium-priced car, the De Soto. The new cars went into production just as Ford retired the Model T, and the pause in Ford production gave Chrysler the opportunity to attract a market for the Plymouth. Chrysler sales grew to 450,543 cars in 1929.[6]

The Airflow Car

In the years that followed, Chrysler introduced a number of improvements to automobiles.[7] The most important innovation was a new car, the Airflow, which Chrysler introduced in 1934 under the Chrysler and De Soto brand names. The new car responded to the fact that by the 1930s paved roads were now more common in the United States and auto companies were in competition with each other to produce cars with stronger engines that could take advantage of these roads to drive at higher speeds. Cars with six- and eight-cylinder engines joined those with four, and cars also began to have closed bodies. Body shape was crucial to the performance of cars at higher speeds. The traction power required to drive an automobile depended on the weight of the car, the frontal area facing the wind, and two coefficients, one representing the resistance of the road surface and one representing the resistance of the air, or drag (sidebar 10.2). The surfaces of paved roads had lower resistance than the surfaces of unpaved ones, but because air pressure increased by the square of velocity, drag increased dramatically at higher speeds. Carl Breer of Chrysler's design team realized that American cars needed more aerodynamically efficient shapes.

Breer began testing new auto body shapes in a wind tunnel during the early 1930s and the Airflow emerged from this research. Weighing about 3,900 pounds, not excessive for the larger automobiles of the time, the Airflow had a streamlined shape that reduced its drag coefficient to about 0.50. Competing automobiles with similar weights and boxier shapes had drag coefficients significantly higher and needed more horsepower to drive at a speed of sixty miles per hour. The new car flowed the windshield into the roof and had a more tapered back. Tapering the back required moving the rear passenger seats forward of the rear axle instead of resting them over it, but the move gave passengers in the back a smoother ride.

· ·

Sidebar 10.2 **The Chrysler Airflow**

The 1934 Chrysler Airflow had an aerodynamically shaped body. The efficiency of the shape can be seen in the traction horsepower of the Airflow compared with that of a contemporary car with a boxier shape, the Packard, at high speed:

C_R	road coefficient	A_F	frontal area of car (in sq. ft.)
W	weight of car (in lbs.)	C_D	drag coefficient
p	air pressure (kV^2) (in lbs. per sq. ft.)	V	velocity (in mph)

$$C_R W + p A_F C_D = T \text{ (traction force)}$$
$$TV/375 = P_T \text{ (traction horsepower)}$$

At a velocity of 60 miles per hour, air pressure p was $(0.00257)(60)^2 = 9.25$ pounds per square foot. Frontal area of both cars is assumed to be 28 square feet. The Airflow required about one-third less horsepower.

1934 Chrysler Airflow   1934 Packard Eight

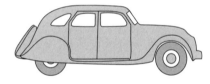   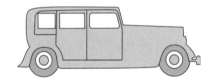

C_R	$W + p$	A_F	C_D		C_R	$W + p$	A_F	C_D
(0.015)	(3974) + (9.25)	(28)	(0.50)		(0.015)	(4640) + (9.25)	(28)	(0.80)

Chrysler Airflow $P_T = \dfrac{TV}{375} = \dfrac{(189)(60)}{375} = 30$ Hp   Packard Eight $P_T = \dfrac{TV}{375} = \dfrac{(277)(60)}{375} = 44$ Hp

Sources: James J. Flink, "The Path of Least Resistance," *Invention and Technology* 5, no. 2 (Fall 1989): 36–39; Beverly Rae Kimes and Henry Austin Clark Jr., *Standard Catalog of American Cars, 1805–1942*, 3rd ed. (Iola, WI: Krause Publications, 1996), pp. 320, 1127.

· ·

Figure 10.2. The Chrysler Airflow. Courtesy of *Automobile Quarterly*, New Albany, IN.

The Airflow was not, however, a commercial success. Moving the rear seating forward made the car more cramped inside, and the aerodynamic shaping of the grille in front of the engine gave the Airflow a bulbous look that critics ridiculed as ungainly and that consumers rejected (figure 10.2).[8] Efforts to redesign the grille more aesthetically could do little to change its basic shape. The Airflow strikingly illustrates the fact that it is not enough to produce a technically efficient design; the design must be socially accepted as well.[9] Ford introduced a streamlined car in 1936, the Lincoln Zephyr, designed by the engineer John Tjaarda. The Zephyr preserved a more traditional looking front and enjoyed a modest success (figure 10.3). General Motors soon brought out more aerodynamically shaped cars of its own. Despite its failure to sell, the Airflow brought streamlined form to American automobiles.

Figure 10.3. The Ford Lincoln Zephyr and the Douglas DC-3. Courtesy of The Henry Ford Museum, Dearborn, MI. Photograph no. 833.66306.

Donald Douglas

Efficient form was an even greater need in aircraft design by the 1930s. The greatest airplane of the 1930s was the streamlined DC-3 of Donald Wills Douglas (1892–1971) (Figure 10.4). Born in Brooklyn, New York, Douglas began a lifelong enthusiasm for aviation when he saw Orville Wright demonstrate his Flyer at Fort Myer, Virginia, in 1909. That autumn, Douglas entered the U.S. Naval Academy in Annapolis, Maryland, but resigned when the Navy opposed his desire to fly. He finished his education at the Massachusetts Institute of Technology, where he roomed with Thomas Edison's

Figure 10.4. Donald W. Douglas. Courtesy of Brown Brothers, Sterling, PA.

son Theodore, whose father offered him a job. Douglas declined and accepted instead an offer to design airplanes for Glenn L. Martin, an early plane manufacturer near Los Angeles, California, who was on contract to build warplanes for the U.S. Army. Douglas soon became known as a talented aircraft designer, and in 1916 he served as chief aeronautical engineer for the Army Signal Corps. But he returned to Martin in 1918 when he failed to win backing for a stronger air force.[10]

Aircraft design after the Wright brothers underwent important changes. The Wright Flyer, with an airframe of all wing and no body, could not carry more than two people and was too fragile to fly at significantly higher speeds. It was also difficult to fly. Airplanes that followed placed a motor in front to pull a longitudinal fuselage crossed by biplane or monoplane wings, with a tail wing and rudder at the end. The pilot sat in a cockpit instead of lying or sitting on the wing, and a single handstick controlled the movement of the plane in pitch, roll, and yaw. Ailerons replaced wing warping. Airframes continued to be made of wood and fabric, and biplane and monoplane wings in the 1910s and 1920s had struts and wires to hold them. Drag did not prevent these aircraft from flying well at speeds under 100 miles per hour.

After World War I, demand for airplanes fell sharply. Donald Douglas wanted to design planes for himself, and despite the market, he gave up his job with Martin to launch his own aircraft company in 1920. Several of his colleagues at Martin joined him and a trickle of contracts kept the business alive until 1923, when the U.S. Navy commissioned Douglas to build a torpedo bomber. The Army then asked him to build four airplanes capable of flying around the world. Douglas produced a biplane that he called the "World Cruiser" identical to his torpedo plane design, except that instead of a bomb load the plane carried extra fuel (figure 10.5). A pilot and a crewman would fly it.

On March 17, 1924, four of the new planes took off from Clover Field near the Douglas assembly building, a former film studio, in Santa Monica, California. The planes flew up to Seattle and then north. One plane went down in Alaska (the crew survived) while the other three proceeded across the Pacific to Japan and then across the mainland of Asia and Europe to Paris, France. Over the North Atlantic a second plane went down but the crew survived and flew a replacement plane from Nova Scotia for the final leg home. On September 3, before 200,000 spectators, the three airplanes returned, two of them after a voyage of 27,550 miles. The 1924 World Cruisers established Douglas as a leading aircraft designer.[11]

Aircraft manufacturing picked up in 1925 when the U.S. Congress authorized aircraft makers to provide airmail service on contract to the Post Office, and airmail planes soon began carrying passengers as well. In 1925 Henry Ford introduced the Ford Trimotor, an innovative aircraft that as a safety feature used three engines, one on each wing and one mounted in front, and also had a metal exterior. A fall in airplane sales in the early 1930s caused Ford to leave the aircraft business.[12] Douglas and other aircraft manufacturers survived during the Depression, but competition to improve the speed and carrying capacity of airplanes challenged designers to innovate.

As aircraft speeds rose in the 1920s, airplane wings encountered more intense drag. In 1922 the French aviator Louis-Charles Bréguet argued that minimizing the ratio of drag to lift by more aerodynamic shaping of the airplane could increase carrying capacity. An aeronautical engineer at Cambridge University, Sir Bennett Melvill Jones,

Figure 10.5. Douglas World Cruisers. Courtesy of the Prints and Photographs Division, Library of Congress, Washington, DC. LC-USZ62-100388.

demonstrated in 1929 that streamlining could greatly reduce the drag caused by flow separation (sidebar 10.1). Airplane designers began to streamline wings and bodies in the 1930s.[13]

Streamlining played a key role in the aircraft produced by William Edward Boeing (1881–1956), who emerged in the 1930s as the principal rival to Donald Douglas. The

Boeing Monomail plane of 1930 pointed to the airplane of the future, with monoplane wings (set under rather than over the fuselage) and with a more streamlined body, retractable landing wheels, and the elimination of wing struts and wires (figure 10.6). But the Boeing plane still had an open cockpit like earlier planes, and it did not carry passengers. The Boeing 247, introduced in 1933, made up for these drawbacks. The 247 was a fully streamlined airplane with all-metal surfaces and an enclosed cockpit and cabin. Its two engines, one on each wing, had casings that flowed into the wing to reduce drag further. Boeing also had a ready customer in its own airline (United Airlines, now an independent carrier). The 247 carried ten passengers and could fly from New York to Los Angeles, with a stop in Cleveland, in twenty hours. But the 247 was not able to turn a profit. Air travelers in the 1930s all flew first class but not enough of them could be carried by the 247 to pay the costs of flying the plane. If aircraft makers were to survive on more than uncertain government contracts, Donald Douglas knew that they had to build airplanes able to carry private passengers and freight at a profit.[14]

The Douglas DC-3

The turning point in commercial aviation was a tragic air crash. Knute Rockne, the legendary football coach at Notre Dame University, died when a foreign-built trimotor crashed in a thunderstorm on March 31, 1931. The airline that flew the plane, Transcontinental and Western (TWA), faced ruin if it could not fly safer planes. The best airplane under development at the time, the planned Boeing 247, had been promised to United Airlines. In a letter sent on August 2, 1932, TWA's vice president, Jack Frye, invited rival aircraft manufacturers to build a better plane. Frye specified that the plane had to carry twelve passengers, fly at an altitude of 21,000 feet, and have a range of 1,000 miles. Such a plane would fly twice as high and twice as far (nonstop) as any airplane in existence in 1932. Donald Douglas accepted the challenge.[15]

In 1915 Congress had established the National Advisory Committee on Aeronautics (NACA), predecessor of today's NASA, to perform aviation research. Its laboratory in Hampton, Virginia, had tested wing shapes, airframes, and other components of aircraft using wind tunnels at windspeeds of several hundred miles per hour. NACA created a

Figure 10.6. The Boeing Monomail airplane. Courtesy of Boeing Image Archives, Chicago, IL. Code 4188. Property of Boeing Management Company. Reprinted under license.

standard series of wing shapes with known performance characteristics that aircraft engineers could use to design airplanes. Douglas combined two of these standard wing shapes, with the help of Jack Northrop, an engineer working with Douglas. The engines were placed one on each wing but close to the fuselage, with NACA-developed cowlings over the engine fronts to reduce drag. The entire body was streamlined (figure 10.7, a and b).[16]

Figure 10.7a. The Douglas DC-1: front view. Courtesy of Boeing Image Archives, Chicago, IL. Negative no. 4983. Property of Boeing Management Company. Reprinted under license.

The new airplane, named the Douglas Commercial or DC-1, made a test flight on July 1, 1933, and nearly crashed. Douglas engineers found that part of the fuel system in the two engines had been mounted backwards![17] By August the plane was flying well, but the model never went into production; by then, Douglas had designed a version with modifications that TWA wanted and the new plane, the DC-2, began service in May 1934. The DC-2 reduced the flight time from New York to Los Angeles from twenty hours to fifteen, carried fourteen passengers, and could travel from coast to coast overnight and not lose a business day. The Boeing 247 could not compete in seating capacity or in speed and soon went out of production.[18]

At the end of 1934, American Airlines asked Douglas to produce a sleeper airplane. Douglas designed the DST or Douglas Skysleeper Transport, a slightly longer version

Figure 10.7b. The Douglas DC-1: rear view. Courtesy of Boeing Image Archives, Chicago, IL. Negative no. 4985. Property of Boeing Management Company. Reprinted under license.

of the DC-2, capable of holding fourteen sleeping berths. These berths could be converted to twenty-one seats for daytime travel. The DST first flew on December 17, 1935, and the daytime version, which received the title DC-3, entered service in 1936. The DC-3 could carry up to twenty-eight passengers and cost only about 10 percent more than the DC-2 to operate. With the new aircraft, airlines could almost double their revenue from each plane and as a result passenger air service finally became profitable. By 1942, of the 322 domestic passenger airplanes in service, 260 were DC-3s. Renamed the C-47 during the Second World War, the DC-3 became the mainstay of wartime U.S. military air transport.[19]

The DC-3 represented an enormous advance in efficiency (sidebars 10.3 and 10.4). The Wright Flyer of 1903 weighed 750 pounds and had a wing surface area of 510

Sidebar 10.3 **The Douglas DC-3**

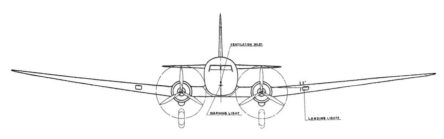

Thrust and Speed

The Douglas DC-3 flew on two Wright Cyclone engines or on two Pratt and Whitney Twin-Wasp engines. The Cyclone engines were rated at 850 horsepower (Hp) at 5,800 feet while the Pratt and Whitney engines were rated at 850 Hp at 8,000 feet (we assume below that 850 Hp was available at 10,000 feet).

The formula $P_T = TV/375$ can be used to calculate thrust in an airplane, with the thrust taking the place of the automobile's traction force (T). If $P_T = 850$ Hp, the thrust of the DC-3 at its cruising speed (V) of 192 miles per hour would have been 1,660 lbs:

$$T = \frac{P(375)}{V} = \frac{(850)(375)}{192} = 1,660 \text{ lbs.}$$

Drag and Lift

At its cruising altitude of 10,000 feet, the air pressure (kV^2) was 0.00189(192)^2 or 69.67 pounds per square foot (rounded to 70 lbs./ft.^{2}). With drag equal to thrust and the surface area of the wingspan (A) equal to 987 square feet, the drag coefficient (C_D) of the DC-3 would have been 0.024:

Drag (thrust)	=	kV^2	A	C_D
1,660 lbs. (approx.)	=	70 lbs./ft.^{2}	987 ft.^{2}	0.024

Fully loaded, the DC-3 weighed 24,000 pounds. With lift equal to weight, and air pressure and surface area the same, the lift coefficient (C_L) would have been 0.347:

Lift (weight)	=	kV^2	A	C_L
24,000 lbs. (approx.)	=	70 lbs. / ft.^{2}	987 ft.^{2}	0.347

Source: C. G. Grey and Leonard Bridgman, eds., *Jane's All the World's Aircraft, 1937* (London: Sampson Low, 1937), pp. 292–94.

Sidebar 10.4 **The Wright Flyer and the Douglas DC-3**

The 1903 Wright Flyer

The 1936 Douglas DC-3

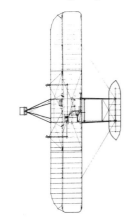

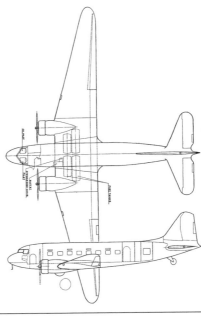

	Wright Flyer	Douglas DC-3
Thrust (T)	84.00 lbs.	1660.00 lbs.
Velocity (V)	30.70 mph	192.00 mph
Thrust Hp (P_T)	6.90 Hp	850.00 Hp
Air Pressure (kV^2)	2.42 lbs./ft.^{2}	70.00 lbs./ft.^{2}
Surface Area (A)	510.00 ft.^{2}	987.00 ft.^{2}
Lift (L)	750.00 lbs.	24,000.00 lbs.
Drag (D)	84.00 lbs.	1660.00 lbs.
Lift Coefficient (C_D)	0.61	0.347
Drag Coefficient (C_D)	0.065	0.024
Lift/Drag	8.93	14.45
Weight/Surface Area	1.41	24.31

Sources: Wright Flyer data from sidebar 6.4, DC-3 data from sidebar 10.3. Some figures rounded. Omega G. East, *Wright Brothers* (Washington, DC: National Park Service, 1961; repr. 1991), p. 39, and Grey and Bridgman, *Jane's All the World's Aircraft, 1937* (London: Sampson Low, 1937), pp. 293.

Figure 10.8. A DC-3 over Chicago. Courtesy of the Still Picture Branch, National Archives, College Park, MD. RG237-G27-1.

square feet, a ratio of weight-to-surface area of 1.41. The DC-3 had a wing surface area of 987 square feet but was far heavier fully loaded at 24,000 pounds. Its ratio of weight-to-surface area, 24,000/987 or 24.31, was a stunning seventeen times that of the Wright Flyer. With a cruising speed of 192 miles per hour, the DC-3 was also more than six times as fast.[20] Douglas designed a machine that was technically outstanding and commercially successful (figure 10.8). The innovation pointed to a future of greater size and

Figure 10.9. The Chrysler Building in New York City. Courtesy of Brown Brothers, Sterling, PA.

Figure 10.10. The Trylon and Perisphere at the 1939–40 New York World's Fair. Courtesy of The Museum of the City of New York. World's Fair Publicity Photos, Box 2.

speed that jet aircraft would eventually realize and that would make air travel a truly mass market.

Streamlining became a leading influence in American art and architecture as well as engineering in the 1930s. A skyscraper in New York City, commissioned by Walter Chrysler, is still a symbol of a streamlined decade with its Art Deco style and tapered top (figure 10.9). Streamlining was the architectural theme of the 1939 New York World's Fair. Americans in the 1930s had endured great privations in the name of efficiency, and the nation looked out on a world preparing for a global war. The Trylon tower and the Perisphere globe at the center of the New York World's Fair, and fair buildings typified by the General Motors pavilion, looked forward to a more hopeful future (figure 10.10).[21]

Streamlining is a metaphor of engineering. In physical terms, a streamlined shape is one that offers the least resistance and the fastest speed with a given amount of power. But the streamlining of the 1930s did not replace the human judgment that made the difference to successful streamlined engineering forms. If efficiency and economy were its only justification, though, streamlining would not have become the symbol of a decade. Streamlined form was also elegant. The best works of modern engineering have combined efficient design, sensitivity to cost, and the vision to realize these two goals where possible in an aesthetic way.

Appendix

The Edison Dynamo and the Parallel Circuit

Perhaps in no other case has the relationship of science to engineering been more significantly misunderstood than in the work of Thomas Edison. Edison continues to be remembered as an inventor who is thought to have merely applied scientific discoveries made by others. In fact, Edison understood the implications of Ohm's and Joule's Laws in a way that eluded some of the leading scientists and theoretically-minded engineers of his time, who believed that the subdivision of electricity for light was fundamentally impossible.

The error on the part of these experts was not faulty scientific logic but sound reasoning—based on narrow assumptions of a kind that are often typical of professional groups when judging new ideas in terms of their dated frames of reference. An example of such thinking, cited in a seminal 1951 article by Harold Passer, occurred in an 1879 book on electric lighting by the British specialist Paget Higgs, who concluded—on the eve of Edison's great breakthrough—that a parallel circuit for distributing electricity to a network of lights could not possibly work.[1] Higgs used detailed calculations based on the latest scientific understanding. The key calculations are as follows:

For a constant voltage, Higgs showed that the maximum power output in an electric circuit consisting of a power source, a line, and lamps occurs when the resistance in the lamps (R_2) is equal to the sum of the resistance in the power source and the resistance in the line (R_1). This equation held with a battery circuit and engineers trying to apply science believed that dynamos had to be designed with R_2 equal to R_1.

For a given voltage V, and $R_2 = \alpha R_1$, the total resistance $R_1 + R_2$ will be $R_1 (1 + \alpha)$, and the current I will be $V / R_1 (1 + \alpha)$. The power in the lamps will then be:

[1] Harold C. Passer, "Electrical Science and the Early Development of the Electrical Manufacturing Industry in the United States," *Annals of Science* [London], vol. 7, no. 4 (December 28, 1951), pp. 382–92; and Paget Higgs, *The Electric Light* (London, 1879), pp. 158–75.

$$I^2 R_2 = \frac{V^2 \alpha R_1}{R_1^2 (1+\alpha)^2} = \frac{V^2}{R_1} \frac{\alpha}{(1+\alpha)^2}.$$

The maximum power in R_2 will be found when the derivative of $\alpha / (1+\alpha)^2$ equals zero:

$$\frac{d}{d\alpha} \frac{\alpha}{(1+\alpha)^2} = 0 = \frac{(1+\alpha)^2 - \alpha(2+2\alpha)}{(1+\alpha)^4},$$

from which $1 + 2\alpha + \alpha^2 - 2\alpha - 2\alpha^2 = 1 - \alpha^2 = 0$, so that $\alpha^2 = 1$, or $\alpha = 1$, from which $R_2 = \alpha R_1 = R_1$.

From this mathematically correct result, Higgs argued that for a series circuit with n lamps, and a constant power source (W_1) to run the dynamo, the heat in the lamps (H) would be:

$$H_1 = W_1 \frac{nR_2}{R_1 + nR_2}$$

He argued next that since the total light varies inversely as n, the light emitted (L_1) in the series circuit would be:

$$L_1 = W_1 \frac{nR_2}{n(R_1 + nR_2)}$$

By the same reasoning, for a parallel circuit with n lamps, the total resistance of the lamps would be R_2/n and thus:

$$L_2 = \frac{R_2 / n}{n(R_1 + R_2 / n)}$$

He then assumed that, for the series circuit, R_1 is so small compared to nR_2 that it can be taken as zero, so that $L_1 = W_1/n$, whereas for the parallel circuit he assumed that R_2/n will be so small compared to R_1 that it can be neglected in the denominator, leading to the result:

$$L_2 = \frac{W_1 R_2}{n^2 R_1}$$

These results were for the networks. Higgs divided each network by n to get the light for each lamp, which was proportional to $1/n^2$ for the series circuit and $1/n^3$ for the parallel one. He concluded that subdividing the current in a parallel circuit was "hopeless."[2]

Higgs, and other experts at the time, made the fundamental error of assuming that the mathematical proof for $R_1 = R_2$ was a reasonable proof for design. As Edison had recognized, it was not. Higgs's result showed the maximum power output, but it did so at a disastrously low efficiency. This can be easily shown for the case where $R_1 = 10$ ohms and $R_2 = 10$ ohms, when compared to the case where $R_1 = 10$ ohms and $R_2 = 90$ ohms.

In both cases, if $V = 100$ volts, then in the first case the current $I = 100/20 = 5$ amps. The dynamo power $P = VI$ will be 100 (5) or 500 watts and the power output to the lamps $P = I^2R$ will be 5^2 (10) or 250 watts, an efficiency of only 50 percent. In the second case, $I = 100/100 = 1$ amp. The dynamo power $P = VI$ will be 100 (1) or 100 watts and the power output to the lamps $P = I^2R$ will be 1^2 (90) or 90 watts, an efficiency of 90 percent. From these relationships Edison recognized the need to reduce resistance in the dynamo and to use a high resistance lamp.

The flaw in Higgs's thinking lay not in his science—he used Ohm's and Joule's Laws correctly—but in his assumption that the engineer should use scientific laws as the basis of design, rather than as guidelines within which to make choices. In engineering, it is only after the designer sees the way to make a true innovation—one that is both efficient and economical—that the equations become useful and essential.

[2] Paget Higgs, *The Electric Light*, p. 174.

Notes

PREFACE

1. *Fortune*, April 5, 2004, p. F1.

2. David P. Billington, *The Innovators: The Engineering Pioneers Who Made America Modern* (New York: John Wiley and Sons, 1996).

3. Harold Evans, with Gail Buckland and David Lefer, *They Made America: From the Steam Engine to the Search Engine: Two Centuries of Innovators* (New York: Little, Brown, 2004).

4. Pauline Maier, Merritt Roe Smith, Alexander Keyssar, and Daniel J. Kevles, *Inventing America* (New York: W. W. Norton, 2002).

5. Scholars have long recognized the independence of engineering innovation. See the symposium papers collected in *Technology and Culture* 17, no.4 (October 1976). Many engineering fields employ people trained primarily in the sciences and their contribution is enormously important. But what matters is what researchers and practitioners do, whether they are engaged primarily in discovery or primarily in design.

6. Walter G. Vincenti, *What Engineers Know and How They Know It: Analytical Studies from Aeronautical History* (Baltimore: Johns Hopkins University Press, 1990), pp. 6–9.

7. Evidence for these statements may be gathered from the experience of the senior author over a quarter century of teaching, and from the experiences of a growing number of younger faculty at public and private institutions around the United States who are adapting the approach to their needs.

CHAPTER ONE

The World's Fairs of 1876 and 1939

1. On the New York fair, see Helen A. Harrison, guest curator, *Dawn of a New Day: The New York World's Fair, 1939/40* (New York: Queens Museum/New York University Press, 1980).

2. For a description of the Philadelphia fair, see Phillip T. Sandhurst, *The Great Centennial Exhibition, 1876* (Philadelphia: P. W. Ziegler, 1876). For the first century of American engineering, see David P. Billington, *The Innovators: The Engineering Pioneers Who Made America Modern* (New York: John Wiley and Sons, 1996).

CHAPTER TWO

Edison, Westinghouse, and Electric Power

1. For the early gaslight industry, see Louis Stotz, with Alexander Jamison, *History of the Gas Industry* (New York: Stettiner Brothers, 1938).

2. For an overview of early electrical discoveries, see Malcolm MacLaren, *The Rise of the Electrical Industry during the Nineteenth Century* (Princeton, NJ: Princeton University Press, 1943), pp. 1–33.

3. On early arc and incandescent light experiments, see Brian Bowers, *A History of Electric Light and Power* (London: Peter Peregrinus, 1992), pp. 64–65.

4. On Faraday's generator, see James Hamilton, *Faraday: The Life* (London: HarperCollins, 2002), pp. 245–51. For the principles of direct-current generators, see Arthur L. Cook and Clifford C. Carr, *Elements of Electrical Engineering: A Textbook of Principles and Practice* (New York: John Wiley and Sons, 1947), pp. 100–153.

5. On the Gramme generator and early arc lighting systems, see MacLaren, *The Rise of the Electrical Industry*, pp. 68–70, 114–20.

6. For Edison's biography, see Paul Israel, *Edison: A Life of Invention* (New York: John Wiley and Sons, 1998). On Menlo Park, see ibid., pp. 119–66, and on the great breakthrough of electric light, pp. 167–90. For a firsthand account, see Francis Jehl, *Menlo Park Reminiscences*, 3 vols. (Dearborn, MI: Edison Institute, 1936–41). Thomas P. Hughes, in *Networks of Power: Electrification in Western Society, 1880–1930* (Baltimore: Johns Hopkins University Press, 1983), pp. 1–46, brings out Edison's integrated approach to the problem of power and light. Israel, *Edison*, pp. 188–90, notes that Edison arrived at a complete system only gradually.

7. On arc light, see Harold C. Passer, *The Electrical Manufacturers: 1875–1900* (Cambridge, MA: Harvard University Press, 1953), pp. 78–83. Through the use of a shunt wire that provided a backup connection, current could flow around an arc light that went out, but the lights on a circuit could not be turned on and off individually.

8. For the opposition to subdividing electricity for light by figures such as Professor Silvanus Thompson and Sir William Preece, see Harold C. Passer, "Electrical Science and the Early Development of the Electrical Manufacturing Industry in the United States," *Annals of Science* (London) 7, no. 4 (December 28, 1951): 383–92.

9. For Edison's power calculations using Ohm's and Joule's laws, see Menlo Park Notebook, No. 10, p. 3, Thomas Edison Papers, Edison National Historic Site, West Orange, New Jersey. For his cost estimates, see Menlo Park Notebook, No. 172, p. 71.

10. The resistance of arc lamps can be calculated from the example of voltage and current given by Passer, *The Electrical Manufacturers*, p. 81.

11. On the availability of a vacuum pump, see ibid., pp. 75–77. For the chemical research that led to a working light bulb, see Byron M. Vanderbilt, *Thomas Edison, Chemist* (Washington, DC: American Chemical Society, 1971), pp. 29–67. For a contemporary report of the breakthrough, see the *New York Herald*, December 21, 1879, pp. 5–6; and for the duration of the bulb, see Israel, *Edison*, p. 186. For his patent, see T. A. Edison, "Electric-Lamp," No. 223,898, United States Patent Office, issued January 27, 1880. Units of candlepower are now called *candelas*.

12. On the different approaches of Edison and Swan, see George Wise, "Swan's Way: A Study in Style," *IEEE Spectrum* 19, no. 4 (April 1982): 66–70.

13. For an overview of the Edison system, see L. H. Latimer, *Incandescent Electric Lighting: A Practical Description of the Edison System* (New York: D. Van Nostrand, 1890). The son of a fugitive slave, Lewis Latimer (1848–1928) drafted the telephone patent illustrations for Alexander Graham Bell and joined Edison in 1884 as an engineer and patent adviser. For Edison's use of feeders and mains and the copper it saved, see Jehl, *Menlo Park Reminiscences*, 2: 820–22. The mains and lamp circuits operated within 2 percent of 110 volts, an acceptable margin.

14. For the Edison generator, see T. A. Edison and Charles T. Porter, "Description of the Edison Steam Dynamo," *Transactions of the American Society of Mechanical Engineers* 3 (1882): 218–25; and Charles L. Clarke, "Edison 'Jumbo' Steam Dynamo," in *"Edisonia": A Brief History of the Early Edison Electric Lighting System* (New York: Association of Edison Illuminating Companies, 1904), pp. 27–59. The armature resistance of Jumbo No. 3, the model used at Pearl Street, was 0.0039 ohms; see Clarke, "Edison 'Jumbo Steam Dynamo," p. 41. On the innovation of low resistance in the dynamo, see Passer, "Electrical Science," pp. 388–91. For expert doubts, see also our appendix.

15. For Edison's cost estimates, see again Menlo Park Notebook, No. 172, p. 71. The actual cost of installing the Pearl Street system was $302,491, of which $151,386 was for the underground

conductors. See Edison Electric Illuminating Company, "Total Cost of Pearl Street Station as Installed," microfilm reel 66, frame 700, Thomas A. Edison Papers Microfilm Edition (Frederick, MD: University Publications of America, 1985–). For the voltage and current of the Pearl Street network, see Jehl, *Menlo Park Reminiscences*, 3: 1077–78. Clarke, "Edison 'Jumbo' Steam Dynamo," pp. 39–41, gives figures for voltage and power that differ only slightly. For Edison's implementation of the system, see Israel, *Edison*, pp. 191–207, and for the funding of Edison's work, see ibid., pp. 173–74, 206. The *watt* as a unit of power began to be used in 1882; Edison in his calculations used the term *energy*. He used the term *webers* to refer to amperes.

16. For Edison's career as an entrepreneur, see André Millard, *Edison and the Business of Innovation* (Baltimore: Johns Hopkins University Press, 1990).

17. On early electric motors, see MacLaren, *The Rise of the Electrical Industry*, pp. 82–106. For the principles of direct-current electric motors, see Cook and Carr, *Elements of Electrical Engineering*, pp. 158–81. For the use of electric power in industry, see Louis C. Hunter and Lynwood Bryant, *A History of Industrial Power in the United States, 1780–1930* (Cambridge, MA: MIT Press, 1991), 3:217–37.

18. For Edison's effort to extend his network using a "three-wire" system, see Jehl, *Menlo Park Reminiscences*, 2:825–27. For the limits of direct-current power transmission, see Cook and Carr, *Elements of Electrical Engineering*, pp. 247–57.

19. For Westinghouse, see Francis E. Leupp, *George Westinghouse: His Life and Achievements* (Boston: Little Brown, 1918). See also the chapters on Westinghouse in Jill Jonnes, *Empires of Light: Edison, Tesla, Westinghouse and the Race to Electrify the World* (New York: Random House, 2003).

20. The article, on the Gaulard and Gibbs alternating-current system, appeared in *Engineering* (London) 39 (May 1, 1885): 158–59. For the analogy of electric transmission to gas pipelines and for the stimulus of the Gaulard and Gibbs system to Westinghouse, see Passer, *The Electrical Manufacturers*, pp. 130–35.

21. For the transformer and alternating-current power transmission, see Cook and Carr, *Elements of Electrical Engineering*, pp. 382–410, 528–37. With alternating current, volt-amps are used instead of watts to denote units of power.

22. See William Stanley, "Alternating-Current Development in America," *Journal of the Franklin Institute* 173 (January–June 1912): 561–80; and on Stanley, see Laurence A. Hawkins, *William Stanley (1858–1916)—His Life and Work* (New York: Newcomen Society, 1951). On the Westinghouse light and power system, see Passer, *The Electrical Manufacturers*, pp. 137–39. In the Lawrenceville test, we assume that Westinghouse used the 50 volt, 2 amp bulbs that he

marketed after the test. We also assume that he sent some additional power to offset line loss during the test, which we and Passer neglect in the calculations given.

23. For the number of A.C. and D.C. central stations in 1890, see Passer, *The Electrical Manufacturers*, p. 150.

24. On the current war, see Jonnes, *Empires of Light*, pp. 141–245. See also Richard Moran, *Executioner's Current: Thomas Edison, George Westinghouse, and the Invention of the Electric Chair* (New York: Alfred A. Knopf, 2002).

25. On the merger of Edison General Electric and Thomson-Houston, see Israel, *Edison*, pp. 332–37.

26. On the Chicago Fair, see Jonnes, *Empires of Light*, pp. 247–75. On the patent-sharing agreement and consolidation of the electrical industry, see Passer, *The Electrical Manufacturers*, pp. 151–64.

27. For case studies of early electric and gas use, see Mark H. Rose, *Cities of Light and Heat: Domesticating Gas and Electricity in Urban America* (University Park, PA: Pennsylvania State University Press, 1995).

28. For the wider influence of electricity in American life, see David E. Nye, *Electrifying America: Social Meanings of a New Technology, 1880–1940* (Cambridge, MA: MIT Press, 1990).

29. On Tesla, see Margaret Cheney, *Tesla: Man Out of Time* (Englewood Cliffs, NJ: Prentice-Hall, 1981). His United States Patent, No. 381,968, was granted on May 1, 1888. For the principle of the induction motor, see Cook and Carr, *Elements of Electrical Engineering*, pp. 415–20.

30. On Steinmetz, see Ronald R. Kline, *Steinmetz: Engineer and Socialist* (Baltimore: Johns Hopkins University Press, 1992). Born with curvature of the spine, Steinmetz overcame great personal as well as professional challenges in his life. For his work on reducing magnetic or "hysteresis" losses, see ibid., pp. 46–52.

31. On Steinmetz's work on alternating-current circuits, see ibid., pp. 77–91.

32. On the tension between Steinmetz and academic engineering, see ibid., pp. 53–61, 105–20. See also Ronald Kline, "Science and Engineering Theory in the Invention and Development of the Induction Motor, 1880–1900," *Technology and Culture* 28, no. 2 (April 1987): 283–313.

33. For the General Electric Laboratory and the work of Coolidge and Langmuir, see Leonard S. Reich, *The Making of Industrial Research: Science and Business at GE and Bell, 1876–1926* (Cambridge: Cambridge University Press, 1985), pp. 62–128.

CHAPTER THREE

Bell and the Telephone

1. For Bell's life, see Robert V. Bruce, *Bell: Alexander Graham Bell and the Conquest of Solitude* (Boston: Little, Brown, 1973).

2. See Alexander Melville Bell, *Visible Speech* (London: Simpkin, Marshall, 1867).

3. For Bell's early interest in electricity and sound, see Bruce, *Bell*, pp. 45–46, 50–51.

4. For Bell's life and work in Boston to 1874, see ibid., pp. 71–103. An invitation to lecture at the Massachusetts Institute of Technology and a tour of the laboratories there stimulated Bell's own interest in research. See ibid., pp. 102–3.

5. For Henry's demonstration, see Joseph Henry, *The Scientific Writings of Joseph Henry*, 2 vols. (Washington, DC: Smithsonian Institution, 1886), 2:434. On Henry's work with electromagnets, see also Albert E. Moyer, *Joseph Henry: The Rise of an American Scientist* (Washington, DC: Smithsonian Institution Press, 1997), pp. 61–77.

6. On Morse and telegraphy, see Carleton Mabee, *The American Leonardo: A Life of Samuel F. Morse* (New York: Alfred A. Knopf, 1943). On the spread of the telegraph, see Robert Luther Thompson, *Wiring a Continent: The History of the Telegraph Industry in the United States, 1832–1866* (Princeton, NJ: Princeton University Press, 1947). With the fortune he made, Cornell founded Cornell University.

7. For the working of simple and multiple telegraphy, see M. D. Fagen, ed., *A History of Engineering and Science in the Bell System*, vol. 1: *The Early Years (1875–1925)* (Basking Ridge, NJ: Bell Telephone Laboratories, 1975), pp. 717–24. On the maturing of the telegraph industry, see also Paul Israel, *From Machine Shop to Industrial Laboratory: Telegraphy and the Changing Context of American Invention, 1830–1920* (Baltimore: Johns Hopkins University Press, 1992).

8. For an account of how his ideas evolved from the harmonic telegraph to the telephone, see Bell's London lecture, October 31, 1877, reproduced in George B. Prescott, *Bell's Electric Speaking Telephone* (New York: D. Appleton, 1884; repr., New York: Arno Press, 1972), pp. 50–82. See also Bruce, *Bell*, pp. 104–11, 120–24; and Michael E. Gorman, Matthew M. Mehalik, W. Bernard Carlson, and Michael Oblon, "Alexander Graham Bell, Elisha Gray and the Speaking Telegraph: A Cognitive Comparison," *History of Technology* 15 (1993): 1–56. Bell's harmonic telegraph patents were No. 161,739 in 1875 and No. 174,465 in 1876.

9. On Hubbard and his aims, see W. Bernard Carlson, "The Telephone as Political Instrument: Gardiner Hubbard and the Formation of the Middle Class in America, 1875–1880," in

Michael Thad Allen and Gabrielle Hecht, eds., *Technologies of Power: Essays in Honor of Thomas Parke Hughes and Agatha Chipley Hughes* (Cambridge, MA: MIT Press, 2001), pp. 25–55.

10. For Bell's partnership with Hubbard and Sanders, see Bruce, *Bell*, pp. 125–29. For his work with Thomas Watson, see ibid., pp. 132–35.

11. For the Reis telephone and telephone research before Bell, see ibid., pp. 117–18.

12. For Bell's meeting with Joseph Henry, see ibid., pp. 139–40.

13. For the breakthrough of June 2, 1875, see ibid., pp. 145–49.

14. On Bell's feelings for Mabel Hubbard and confrontation with her father, see ibid., pp. 151–61.

15. Alexander Graham Bell, "Improvement in Telegraphy," Letters Patent No. 174,465, March 7, 1876, United States Patent Office, Washington, DC.

16. On Elisha Gray and his telephone, see David A. Hounshell, "Elisha Gray and the Telephone: On the Disadvantages of Being an Expert," *Technology and Culture* 16, no. 2 (April 1975): 133–61; and Gorman et al., "Alexander Graham Bell, Elisha Gray and the Speaking Telegraph: A Cognitive Comparison," pp. 1–56.

17. For Bell's knowledge of variable resistance, see Bruce, *Bell*, pp. 144–45. Questions later arose over whether the inclusion of variable resistance in Bell's patent claim was the result of his attorneys having obtained advance knowledge of Gray's caveat and whether Bell's patent application was moved ahead of Gray's caveat on the day that both were filed in Washington. See A. Edward Evenson, *The Telephone Patent Conspiracy of 1876: The Elisha Gray–Alexander Bell Controversy and Its Many Players* (Jefferson, NC: McFarland, 2000). These charges have never been proved. See Bruce, *Bell*, pp. 167–73.

18. On Bell's liquid telephone experiments, see Bruce, *Bell*, pp. 177–81. Bell's own patent was not for a liquid telephone. He built and tested one after his attorneys in Washington obtained a copy of Gray's caveat.

19. Alexander Graham Bell, "Telephone," Letters Patent No. 186,787, January 30, 1877, United States Patent Office, Washington, DC.

20. On Edison's telephone work, see Paul Israel, *Edison: A Life of Invention* (New York: John Wiley and Sons, 1998), pp. 130–41.

21. On the Western Union litigation, see Bruce, *Bell*, pp. 260–80.

22. On Bell's later life, see ibid., esp. p. 231.

23. On the networking of early telephones, see *A History of Engineering and Science in the Bell System*, 1:468–73. AT&T absorbed American Bell on December 30, 1899; see ibid., pp. 34–35.

See also George David Smith, *Anatomy of a Business Strategy: Bell, Western Electric, and the Origins of the American Telephone Industry* (Baltimore: Johns Hopkins University Press, 1985). On Vail, see John Brooks, *Telephone: The First Hundred Years* (New York: Harper & Row, 1976), pp. 74–85.

24. On the technical problems of long-distance transmission, see A *History of Engineering and Science in the Bell System*, 1:233–34. Copper costs for exposed wire lines increased as the square of distance and for cable (covered) lines increased as the cube of distance.

25. On the Bell System's reliance on engineers with more advanced training, see Leonard S. Reich, *The Making of American Industrial Research: Science and Industry at GE and Bell, 1876–1926* (Cambridge: Cambridge University Press, 1985), pp. 142–50. See also Neil H. Wasserman, *From Invention to Innovation: Long-Distance Telephone Transmission at the Turn of the Century* (Baltimore: Johns Hopkins University Press, 1985), pp. 20–33. The new theoretical research was the work of James Clerk Maxwell and Oliver Heaviside.

26. On the development of inductive loading and its results, see A *History of Engineering and Science in the Bell System*, 1:241–53; and for Campbell's role, see Wasserman, *From Invention to Innovation*, pp. 40–53, 67–90. Campbell spaced the coils at distances determined by the average wavelength of voice frequencies.

27. On Pupin and the patents, see Wasserman, *From Invention to Innovation*, pp. 90–100; and James E. Brittain, "The Introduction of the Loading Coil: George A. Campbell and Michael I. Pupin," *Technology and Culture* 11, no. 1 (January 1970): 36–57.

28. For the development of amplification and transcontinental telephony, see A *History of Engineering and Science in the Bell System*, 1:252–77. See also Lillian Hoddeson, "The Emergence of Basic Research in the Bell Telephone System, 1875–1915," *Technology and Culture* 22, no. 3 (July 1981): 512–44. For the call by Bell and Watson, see Bruce, *Bell*, p. 482.

29. On frequency multiplexing in telephony and Campbell's invention of the wave filter, see A *History of Engineering and Science in the Bell System*, 1:277–90, 899.

30. For the growth of the telephone network and its impact and interaction with society, see again, Brooks, *Telephone*; and Claude S. Fischer, *America Calling: A Social History of the Telephone to 1940* (Berkeley: University of California Press, 1992). For its international influence, see Peter Young, *Person to Person: The International Impact of the Telephone* (Cambridge: Granta Editions, 1991). For AT&T's 1913 divestiture of Western Union and agreement with the federal government, see Brooks, *Telephone*, pp. 132–37.

31. For Bell's involvement with science and scientists, see Bruce, *Bell*, pp. 369–78.

32. See Hounshell, "Elisha Gray and the Telephone," pp. 152–61.

CHAPTER FOUR

Burton, Houdry, and the Refining of Oil

1. For the American steel industry in the late nineteenth and early twentieth centuries, see Thomas J. Misa, *A Nation of Steel* (Baltimore: Johns Hopkins University Press, 1995). On the decline of the industry, see ibid., pp. 253–82.

2. On early illuminants, see Harold F. Williamson and Arnold R. Daum, *The American Petroleum Industry: The Age of Illumination, 1859–1899* (Evanston, IL: Northwestern University Press, 1959), pp. 27–60. This book, together with its companion volume, Harold F. Williamson, Ralph L. Andreano, Arnold R. Daum, and Gilbert C. Klose, *The American Petroleum Industry: The Age of Energy, 1899–1959* (Evanston, IL: Northwestern University Press, 1959), provides a technical history of the oil industry. For a more recent account that places the industry in a larger historical context, see Daniel Yergin, *The Prize: The Epic Quest for Oil, Money, and Power* (New York: Simon and Schuster, 1991). For a description of refining processes, see also William L. Leffler, *Petroleum Refining in Nontechnical Language*, 3rd ed. (Tulsa, OK: PennWell, 2000).

3. B. Silliman Jr., *Report on the Rock Oil, or Petroleum, from Venango Co., Pennsylvania, with Special Reference to Its Use for Illumination and Other Purposes* (New Haven, CT: J. H. Benham's Steam Power Press, 1855), p. 15.

4. On Drake and the early oil industry in Pennsylvania, see Williamson and Daum, *The Age of Illumination*, pp. 63–114.

5. On early refining, see ibid., pp. 202–27. See also James G. Speight, *The Chemistry and Technology of Petroleum* (New York: Marcel Dekker, 1991), pp. 314, 506–10. Earlier sources give the boiling range of kerosene as 350 to 525 degrees Fahrenheit (175–275 degrees Celsius). See William A. Gruse and Donald R. Stevens, *Chemical Technology of Petroleum* (New York: McGraw-Hill, 1960), p. 487. The description of refining is simplified here.

6. On Rockefeller's life, see Ron Chernow, *Titan: The Life of John D. Rockefeller, Sr.* (New York: Random House, 1998).

7. On the consolidation of the oil refining industry under Standard Oil, see Williamson and Daum, *The Age of Illumination*, pp. 301–8, 343–462. Production data are from ibid., pp. 471, 485. On the consolidation of Standard Oil, see also Chernow, *Titan*, pp. 129–82, 197–215.

8. For the retail prices of kerosene, see Williamson and Daum, *The Age of Illumination*, pp. 247, 680.

9. On the 1882 trust agreement and subsequent use of Standard of New Jersey as a holding company, see ibid., pp. 466–70, 715; and Chernow, *Titan*, pp. 224–29, 429–30.

10. Tarbell's articles were published in book form as Ida Tarbell, *The History of the Standard Oil Company*, 2 vols. (New York: McClure, Phillips, 1904–5).

11. On the circumstances and litigation leading to the 1911 breakup of Standard Oil, see Williamson et al., *The Age of Energy, 1899–1959*, pp. 4–14; and Chernow, *Titan*, pp. 537–59. Williamson emphasizes changes in the industry that lessened Standard Oil's market dominance before 1911, while Chernow underlines how the company's arrogance and miscalculations contributed to its breakup.

12. For the discovery and problems of Lima-Indiana oil, see Williamson and Daum, *The Age of Illumination*, pp. 589–613. The Rockefeller quotation is from Nevins, *John D. Rockefeller: The Heroic Age of American Enterprise*, vol. 2 (New York: Charles Scribner's Sons, 1940), p. 7.

13. For information on Herman Frasch, see his entry in the *National Cyclopaedia of American Biography* (New York: James T. White, 1926), 19:347–48. On the Frasch process, see also Chernow, *Titan*, pp. 284–88.

14. For oil production data to 1899, see Williamson and Daum, *The Age of Illumination*, pp. 614–15. The figure for 1885 may be a yearly average for 1883–85.

15. On the kerosene yield of crude oil, see ibid., p. 208; and William M. Burton, "Medal Address: Chemistry in the Petroleum Industry," *Journal of Industrial and Engineering Chemistry* 10, no. 6 (June 1918): 485. For the percentage of gasoline produced by simple distillation, see Paul H. Giddens, *Standard Oil Company (Indiana): Oil Pioneer of the Middle West* (New York: Appleton-Century-Crofts, 1955), p. 140; and Speight, *Chemistry and Technology of Petroleum*, p. 529.

16. On Burton's life, see his entry in the *National Cyclopaedia of American Biography*, 41:41. Biographical notices on Humphreys and Rogers have not been found.

17. On the development and results of the Burton process, see John Lawrence Enos, *Petroleum Progress and Profits: A History of Process Innovation* (Cambridge, MA: MIT Press, 1962), pp. 1–59. On thermal cracking, see also Speight, *Chemistry and Technology of Petroleum*, pp. 529–44.

18. Burton, "Medal Address," p. 485.

19. Ibid.

20. On the efficiency of thermal cracking, raising production of gasoline per barrel of crude oil to about 40 percent, see Giddens, *Standard Oil Company (Indiana)*, pp. 140, 146, 148.

21. Quotation is from ibid., p. 149.

22. W. M. Burton, "Manufacture of Gasolene," United States Patent No. 1,049,667, January 7, 1913. The patent was owned by the company and taken in Burton's name.

23. Burton, "Medal Address," p. 485.

24. Walter G. Vincenti, "Control Volume Analysis: A Difference in Thinking between Engineering and Physics," *Technology and Culture* 23, no. 2 (April 1982): 145–74. Quotation is from p. 166.

25. Burton, "Medal Address," p. 485.

26. For gasoline demand in 1899 and 1919, and the proportions of refined oil and cracked gasoline from 1913 to 1929, see Williamson et al., *The Age of Energy*, pp. 146, 192–95, and 395.

27. For the rival processes, see Williamson et al., *The Age of Energy*, pp. 159–62, 376–85. On the Dubbs process, see ibid., pp. 154–59.

28. On knocking and antiknock additives, see Gruse and Stevens, *Chemical Technology of Petroleum*, pp. 439–69. On Thomas Midgely, see G. B. Kauffman, "Midgely: Saint or Serpent?" *Chemtech* 19, no. 12 (December, 1989): 717–25. Lead was removed in the 1970s because it interfered with catalytic converters, which were installed to control automobile emissions.

29. On octane ratings, see Burnett, *Gasoline: From Unwanted By-product to Essential Fuel for the Twentieth Century* (Stony Brook, NY: NLA Monograph Series, 1991), pp. 66–69. The proper name for iso-octane is 2,2-dimethylpentane. The term iso-octane is used here for simplicity. The octane rating was the average of two numbers: the motor octane number (MON) and the research octane number (RON), obtained from tests.

30. For the octane of gasoline in 1930, see Gruse and Stevens, *Chemical Technology of Petroleum*, p. 438.

31. On Houdry's life, see Alex G. Oblad, "The Contributions of Eugene J. Houdry to the Development of Catalytic Cracking," in Burton H. Davis and William P. Hettinger Jr., eds., *Heterogeneous Catalysis: Selected American Histories*, ACS Symposium Series 222 (Washington, DC, American Chemical Society, 1983), pp. 612–20.

32. On Houdry's work in catalytic cracking, see ibid., and Enos, *Petroleum Progress and Profits*, pp. 131–62. On catalytic cracking, see also Speight, *Chemistry and Technology of Petroleum*, pp. 545–51.

33. On the efficiency of catalytic cracking, see Gruse and Stevens, *Chemical Technology of Petroleum*, pp. 373–75.

34. For example, among the hexanes produced by straight-run, thermal cracking, and catalytic cracking, the proportions of branched hexane molecules were 49, 37, and 91 percent, respectively. See ibid., p. 77. By 1944, there were twenty-nine catalytic cracking units in the United States that produced almost all of the aviation fuel used by the Allies in World War II. See Williamson et al., *The Age of Energy*, p. 620.

35. On the different processes of catalytic cracking, see Williamson et al., *The Age of Energy*, pp. 620–26. On reforming and alkylation, see ibid., pp. 626–33.

CHAPTER FIVE

Ford, Sloan, and the Automobile

1. For automobile registrations in 1900 and 1939, see *Automobile Facts and Figures, 1940* (Detroit: Automobile Manufacturers Association, 1940), p. 11. These numbers do not include trucks.

2. On the Otto engine, see Lynwood Bryant, "The Silent Otto," *Technology and Culture* 7, no. 2 (1966): 184–200; and "Origin of the Four-Stroke Cycle," *Technology and Culture* 8, no. 2 (1967): 178–98.

3. For the European origins of automobile engineering, see James M. Laux, *The European Auto Industry* (New York: Twayne Publishers, 1992), pp. 1–50. On the Duryea car, see Richard P. Scharchburg, *Carriages without Horses: J. Frank Duryea and the Birth of the American Automobile Industry* (Warrendale, PA: Society of Automotive Engineers, 1993).

4. On early steam cars, see George S. May, "Stanley Motor Carriage Company," and James M. Laux, "Steam Cars" and "White Motor Company," in George S. May, ed., *Encyclopedia of American Business History and Biography: The Automobile Industry, 1896–1920* (New York: Bruccoli Clark Layman and Facts on File, 1990), pp. 423–27, 452–57. For engineering arguments on behalf of steam cars, see the papers by Abner Doble, John Sturgess, and Prescott Warren, together with discussion by others, in *Transactions of the Society of Automotive Engineers* 13, part 1 (1918): 338–86.

5. For the engineering of early electrics, see C. E. Woods, *The Electric Automobile: Its Construction, Care, and Operation* (Chicago: Herbert S. Stone, 1900). For a comparison of steam, electric, and gasoline car performance, see Ernest Henry Wakefield, *History of the Electric Automobile: Battery-Only Powered Cars* (Warrendale, PA: Society of Automotive Engineers, 1994), p. 37. For the social history of electric cars, see Michael Brian Schiffer, *Taking Charge: The Electric Automobile in America* (Washington, DC: Smithsonian Institution Press, 1994); and on the electric self-starter, see Stuart Leslie, *Boss Kettering* (New York: Columbia University Press, 1983), pp. 46–50.

6. On early European luxury cars, see Laux, *The European Auto Industry*, pp. 1–50. On the Pierce-Arrow company, see Beverly Rae Kimes, "The Pierce-Arrow Motor Car Company," in May, *Encyclopedia of American Business History and Biography: The Automobile Industry, 1896–1920*, pp. 389–91.

7. The classic biography of Henry Ford is Allan Nevins with Frank Ernest Hill, *Ford: The Times, the Man, the Company*; Allan Nevins and Frank Ernest Hill, *Ford: Expansion and*

Challenge 1915–1933; and Allan Nevins and Frank Ernest Hill, *Ford: Decline and Rebirth 1933–1962* (New York: Charles Scribner's and Sons, 1954–63). For a recent biography, see Douglas Brinkley, *Wheels for the World: Henry Ford, His Company, and a Century of Progress* (New York: Viking Books, 2003). On Ford's early life, the quadricycle, his meeting with Edison, and his first company, see Brinkley, *Wheels for the World*, pp. 1–37.

8. For an account of the Fulton race, see David P. Billington, *The Innovators: The Engineering Pioneers Who Made America Modern* (New York: John W. Wiley and Sons, 1996), pp. 42–44. On the Duryea race, see Scharchburg, *Carriages without Horses*, pp. 121–25.

9. On Ford's racing career and second company, see Nevins, *Ford: The Times, the Man, the Company*, pp. 192–219; and Brinkley, *Wheels for the World*, pp. 37–48.

10. On Ford's third company, see Nevins, *Ford: The Times, the Man, the Company*, pp. 213–332; and Brinkley, *Wheels for the World*, pp. 49–89.

11. On the Ford Model N, see Nevins, *Ford: The Times, the Man, the Company*, pp. 323–38. The Ford 1906–7 figures included Models R, S, and K.

12. On the Ford Model T, see Nevins, *Ford: The Times, the Man, the Company*, pp. 387–93; and Brinkley, *Wheels for the World*, pp. 99–106. Henry Ford gives his business and engineering reasons for developing the Model T in his book, *My Life and Work* (Garden City, NY: Doubleday, Page, 1922), pp. 33–76. The book was written in collaboration with Samuel Crowther.

13. For the engineering of the Model T, see the *Ford Manual: For Owners and Operators of Ford Cars* (Detroit: Ford Motor Company, 1914).

14. For Watt's horsepower formula, see John Farey, *A Treatise on the Steam Engine: Historical Practical and Descriptive* (London: Longman, Rees, Orme, Brown, and Green, 1827), pp. 438–40. Watt estimated that a horse could lift 330 pounds 100 feet in one minute, or 33,000 foot-pounds per minute. He divided *PLAN* by 33,000 to calculate the indicated horsepower of a reciprocating engine. For measurements of engine power at the time, see Robert L. Streeter, *Internal Combustion Engines* (New York: McGraw-Hill, 1923), pp. 247–60. These formulas are still employed today. See John B. Heywood, *Internal Combustion Engine Fundamentals* (New York: McGraw-Hill, 1988), pp. 49–51. The values of *LAN* and Hp are drawn from Nevins, *Ford: The Times, the Man, the Company*, pp. 387–93; and Floyd Clymer, *Henry's Wonderful Model T, 1908–1927* (New York: McGraw-Hill, 1955), p. 127 (fig. 20). Average pressure *P* is inferred. Ford's method of calculating horsepower in the Model T was $D^2 n/2.5$, where *D* is the diameter of the wheel in feet, *n* is the number of crankshaft revolutions per minute, and 2.5 was an empirical number. See the *Ford Manual*, p. 30. With $D = 3.75$ and $n = 4$, Ford's formula gave the Model T a horsepower of 22.5.

15. To calculate brake horsepower, engineers multiply the speed of the crankshaft (N_C), in revolutions per minute, by the turning force or *torque* (T_Q) of the crankshaft, measured in foot-pounds using a meter called a brake (brake horsepower is named for this kind of brake, not for the car's brakes). This result is multiplied by 2π and divided by 33,000 foot-pounds per minute. For the Model T at about 37 miles per hour, brake hp (P_B) = $T_Q N_C 2\pi/33,000 = (70)(1500)(6.28)/33,000 = 19.98$ Hp. Brake hp was thus 85 percent of indicated hp. Traction horsepower (P_T) in the model T was about 60 percent of brake hp, or $P_T = P_B(0.60) = 12$ hp.

16. Engineers today also refer to traction horsepower as the road-load power.

17. On gearing systems in cars, see Philip G. Gott, *Changing Gears: The Development of the Automotive Transmission* (Warrendale, PA: Society of Automotive Engineers, 1991).

18. For a description of Ford mass production, see Horace Lucien Arnold and Fay Leone Faurote, *Ford Methods and the Ford Shops* (New York: Engineering Magazine Company, 1915). See also David A. Hounshell, *From the American System to Mass Production, 1800–1932: The Development of Manufacturing Technology in the United States* (Baltimore: Johns Hopkins University Press, 1984), pp. 217–61.

19. For the prices of Ford cars from 1908/9–1916/17, see Clymer, *Henry's Wonderful Model T*, pp. 109–21. For Model T production figures, see ibid., p. 134. Ford sales data were based on a fiscal year, not a calendar year, but Ford's share of the market may be estimated from these figures, using for comparison the historical data for annual U.S. automobile production in *Automobile Facts and Figures, 1940*, p. 5.

20. On the Selden patent controversy, see William Greenleaf, *Monopoly on Wheels: Henry Ford and the Selden Automobile Patent* (Detroit: Wayne State University Press, 1961).

21. On the five-dollar day and Ford's relations with labor, see Stephen Meyer, *The Five Dollar Day: Labor Management and Social Control in the Ford Motor Company, 1908–1921* (Albany: State University of New York Press, 1981). Ford tried to regulate the private lives of his workers through his company's Sociological Department. These efforts helped some but were regarded by many as intrusive. Ford's hostility to labor unions was shared by other manufacturers of his time, and Ford was the last of the big three automakers to accept unionization of his work force, after a strike, in 1941.

22. On Durant and the early history of General Motors, see Bernard A. Weisberger, *The Dream Maker: William C. Durant, Founder of General Motors* (Boston: Little Brown, 1979); and Axel Madsen, *The Deal Maker: How William C. Durant Made General Motors* (New York: John Wiley and Sons, 1999).

23. On Sloan's management of General Motors, see Alfred P. Sloan Jr., *My Years with General Motors*, ed. John McDonald with Catherine Stevens (New York: Doubleday, 1964), pp. 45–70, 149–68, 238–47. See also Arthur J. Kuhn, *GM Passes Ford, 1918–1938: Designing the General Motors Performance-Control System* (University Park, PA: Pennsylvania State University Press, 1986); and David R. Farber, *Sloan Rules: Alfred P. Sloan and the Triumph of General Motors* (Chicago: University of Chicago Press, 2002).

24. On the decline of the Model T in the 1920s, see Nevins and Hill, *Ford: Expansion and Challenge, 1915–1933*, pp. 379–436; and Sloan, *My Years with General Motors*, pp. 158–62.

25. On Ford's view of business, see Ford, *My Life and Work*, pp. 37–39.

26. On Ford's view of history, see John B. Rae, *Great Lives Observed: Henry Ford* (Englewood Cliffs, NJ: Prentice-Hall, 1969), pp. 53–54. For his anti-Semitism, see Albert Lee, *Henry Ford and the Jews* (New York: Stein and Day, 1980). On the Ford Foundation, see Richard Magat, *The Ford Foundation at Work: Philanthropic Choices, Methods, and Styles* (New York: Plenum Press, 1979).

27. On the air-cooled car (also called copper-cooled, since copper plating inside cooled the engine), see Stuart W. Leslie, "Charles F. Kettering and the Copper-Cooled Engine," *Technology and Culture* 20, no. 4 (October 1979): 752–76.

28. Sloan, *My Years with General Motors*, pp. 71–96.

29. On the adding of lead to gasoline, see Herbert L. Needleman, "Clamped in a Straitjacket: The Insertion of Lead into Gasoline," *Environmental Research* 74, no. 2 (August 1997): 95–103; and Farber, *Sloan Rules*, pp. 81–86.

30. Frederick Winslow Taylor, *The Principles of Scientific Management* (New York: Harper and Brothers, 1911). For Taylor and his influence, see Robert Kanigel, *The One Best Way: Frederick Winslow Taylor and the Enigma of Efficiency* (New York: 1997). For the misguided idea that modern technology is a system crushing everything into a "one best way," see Jacques Ellul, *The Technological System* (New York: Knopf, 1964).

31. On the difference between Taylor and Ford, see Hounshell, *From the American System to Mass Production*, pp. 249–53.

32. See Hounshell, *From the American System to Mass Production*, pp. 252–53. Recent scholarship has questioned whether technology prescribed a single predetermined path for the automobile industry. See Michel Freyssenet et al., *One Best Way? Trajectories and Industrial Models of the World's Automobile Producers* (New York: Oxford University Press, 1998).

33. For the impact of the automobile, see James J. Flink, *The Automobile Age* (Cambridge, MA: MIT Press, 1988); Rudi Volti, "A Century of Automobility," and Ronald R. Kline and Trevor J. Pinch, "Users as Agents of Technological Change: The Social Construction of the Automobile in the Rural United States," both in *Technology and Culture* 37, no. 4 (October 1996): 663–85, and 763–95.

CHAPTER SIX

The Wright Brothers and the Airplane

1. On early aviation, see John D. Anderson Jr., *A History of Aerodynamics and Its Impact on Flying Machines* (Cambridge: Cambridge University Press, 1997), pp. 14–62. For the change brought by the industrial revolution, see Tom D. Crouch, "Aeronautics in the Pre-Wright Era: Engineers and the Airplane," in Richard P. Hallion, ed., *The Wright Brothers: Heirs of Prometheus* (Washington, DC: Smithsonian Institution Press, 1985), pp. 3–19.

2. For Cayley's work, see Charles H. Gibbs-Smith, *Sir George Cayley's Aeronautics, 1796–1855* (London: Her Majesty's Stationary Office, 1962); and Anderson, *A History of Aerodynamics*, pp. 62–80. On the basic principles of aerodynamics, see Anderson, *A History of Aerodynamics*, pp. 3–11. The direction of lift is relative to the direction of the wind.

3. Clement Ader of France, Sir Hiram Maxim of England, and others made hops in the late nineteenth century with airplanes powered by steam engines but did not achieve steady level flight. See Charles H. Gibbs-Smith, *Aviation: An Historical Survey from Its Origins to the End of World War II* (London: Her Majesty's Stationary Office, 1985), pp. 59–63. On the failure of aerodynamic theory to inform practical efforts to fly in the second half of the nineteenth century, see Anderson, *A History of Aerodynamics*, pp. 114–19.

4. See Otto Lilienthal, *Birdflight as the Basis of Aviation*, trans. A. W. Isenthal (New York: Longmans Green, 1911; repr., Hummelstown, PA: Markowski International, 2001); and Anderson, *A History of Aerodynamics*, pp. 138–64. For Lilienthal's influence on American aviation, see Tom D. Crouch, *A Dream of Wings: Americans and the Airplane, 1875–1905* (Washington, DC: Smithsonian Institution Press, 1989), pp. 157–74.

5. For an account of Langley's research and model test flights through 1896, with Bell's testimony, see S. P. Langley, "The 'Flying Machine,'" *McClure's Magazine* 9, no. 2 (June 1897): 647–60.

6. For a review of Langley's work, completed by Manly, see Samuel P. Langley and Charles M. Manly, *Langley Memoir on Mechanical Flight*, Smithsonian Contributions to Knowledge, vol. 27, no. 3 (Washington, DC: Smithsonian Institution, 1911). See also Anderson, *A History of Aerodynamics*, pp. 164–92.

7. For a biography of the Wright brothers, see Tom Crouch, *The Bishop's Boys: A Life of Wilbur and Orville Wright* (New York: W. W. Norton, 1989).

8. Octave Chanute, *Progress in Flying Machines* (New York: M. N. Forney, 1894; repr. Mineola, NY: Dover Publications, 1997). On Chanute and his research, see Crouch, *A Dream of Wings*, pp. 175–202.

9. The Wright brothers explained their basic ideas in Orville Wright and Wilbur Wright, "The Wright Brothers' Aëroplane," *Century Magazine* 76, no. 5 (September 1908): 641–50. For an examination of their work, see Anderson, *A History of Aerodynamics*, pp. 201–43. On their early interest in aviation, see Crouch, *The Bishop's Boys*, pp. 157–80.

10. See again the basic principles of flight in Anderson, *A History of Aerodynamics*, pp. 3–11.

11. On Alphonse Pénaud, see Gibbs-Smith, *Aviation*, pp. 43–44.

12. For the influence of bicycles in the thinking of the Wright brothers, see Tom Crouch, "How the Bicycle Took Wings," *American Heritage of Invention and Technology* 2, no. 1 (Summer 1986): 11–16. For the 1899 kite experiments, see Orville Wright's account in Marvin W. McFarland, ed., *The Papers of Wilbur and Orville Wright*, 2 vols. (New York: McGraw-Hill, 1953), 1: 5–12.

13. For a report on their glider tests, see Wilbur Wright, "Experiments and Observation in Soaring Flight," *Journal of the Western Society of Engineers* 8 (August 1908): 400–417. On the 1900 glider tests at Kitty Hawk, see also Crouch, *The Bishop's Boys*, pp. 181–99; and Peter L. Jakab, *Visions of a Flying Machine: The Wright Brothers and the Process of Innovation* (Washington, DC: Smithsonian Institution Press, 1990), pp. 83–101.

14. For the dimensions and construction of the first glider, see Jakab, *Visions of a Flying Machine*, pp. 93–94. The highest point of curvature on the top surface was close to the front edge, which helped stabilize the wing against shifts in the center of pressure underneath. See ibid., pp. 67–68. The camber ratio of the 1900 glider wing was 1:22. In a stall, an airplane with main and tail wings would lose lift in the main wings slightly before losing it in the tail wings. Placing the tail wings in front of the main wings caused the former to stall first. The main wings still had some lift, giving the plane a flatter descent, like a parachute. See ibid., pp. 68–72. At the higher speeds achieved by later aircraft, this "canard" configuration proved impractical.

15. On these troubling results, see ibid., pp. 100–101.

16. For the second glider and its difficulties, see ibid., pp. 102–14. On the 1901 tests, see also Crouch, *The Bishop's Boys*, pp. 200–213. The 1901 plane had a camber ratio of 1:12 to start, which the Wrights reduced to 1:19 after the trouble with pitch became apparent.

17. For the published version of his presentation, see Wilbur Wright, "Some Aeronautical Experiments," *Journal of the Western Society of Engineers* 6 (December 1901): 489–510.

18. For the formulas and numbers used by the Wright brothers to design their gliders, see McFarland, *The Papers of Wilbur and Orville Wright*, 1: 572–77. The term k is obtained from the formula $kV^2 = 1/2\rho V^2$, where ρ is the density of air at sea level in pounds-seconds squared per foot to the fourth power, and V is in feet per second. For k where V is in miles per hour, we multiply $1/2\rho V^2$ by $(5280/3600)^2 = (1.47)^2$. Since the air density at sea level is 0.002377, the correct value for k there is $1/2(0.002377)(1.47)^2 = 0.00257$, not the 0.005 value of Smeaton's coefficient.

19. On the wind tunnel tests, see ibid., 1:577–93. See also Anderson, *History of Aerodynamics*, pp. 216–35; and Jakab, *Visions of a Flying Machine*, pp. 115–59.

20. See Anderson, *History of Aerodynamics*, pp. 168–69, 209–10. In 1891 Langley realized that Smeaton's coefficient was inaccurate and gave it a value of 0.0033. The Wrights seem to have arrived at their number independently.

21. On the the improved wing design, see Jakab, *Visions of a Flying Machine*, pp. 150–54, 158–59.

22. For the successful 1902 glider tests, see ibid., pp. 163–82; and Crouch, *The Bishop's Boys*, pp. 229–41.

23. For the lift and drag coefficients of wing surface no. 12, the shape chosen by the Wright brothers for their wings, see McFarland, ed., *The Papers of Wilbur and Orville Wright*, 1:579, 583. On the design specifications of the 1903 Wright Flyer, see Howard S. Wolko, "Structural Design of the 1903 Wright Flyer," in Howard S. Wolko, ed., *The Wright Flyer: An Engineering Perspective* (Washington, DC: Smithsonian Institution Press, 1987), pp. 97–106.

24. On the Wright engine, see Harvey H. Lippincott, "Propulsion Systems of the Wright Brothers," in Wolko, *The Wright Flyer*, pp. 82–95. The engine weight is unclear but was between 140 and 179 pounds. See Jakab, *Visions of a Flying Machine*, p. 192.

25. Wright and Wright, "The Wright Brothers' Aëroplane," p. 648.

26. On the propeller research of the Wright brothers, see McFarland, ed., *The Papers of Wilbur and Orville Wright*, 1:594–640; and Lippincott, "Propulsion Systems of the Wright Brothers," pp. 79–82. The brothers positioned two propeller blades behind the wings and had the blades rotate in opposite directions to prevent them from turning the plane left or right on its yaw axis. It is not clear that the Wrights realized, after estimating the power loss in the propellers,

that their original engine power estimate of 8.4 horsepower was insufficient for the worst case of having to fly at thirty-five miles per hour.

27. For the final success at Kitty Hawk in 1903, see Crouch, *The Bishop's Boys*, pp. 253–61, 263–72; and Jakab, *Visions of a Flying Machine*, pp. 183–212.

28. Langley and Manly, *Langley Memoir on Mechanical Flight*, pp. 207–17. Langley and Manly relied on a *dihedral* or slight upward angle of the wing tips to afford some stability against unwanted roll. The Wright Flyer's wing tips actually drooped slightly but the Wrights relied on manual controls (wing warping) to stabilize the plane in roll.

29. On Langley's climactic failure, see ibid., pp. 255–81; and Crouch, *A Dream of Wings*, pp. 255–93. Langley insisted on launching over water out of concern for pilot safety. Unwilling to allow his airplane to be patented for private profit, Langley refused offers of private funding to carry on his work.

30. Orville Wright, "How We Made the First Flight," in Hallion, *The Wright Brothers*, pp. 101–9. Quotation is from pp. 107–8. Originally printed in the magazine *Flying* (December 1913).

31. Estimates of windspeed that morning varied from twenty to twenty-seven miles per hour. See McFarland, *The Papers of Wilbur and Orville Wright*, 1:395n. For Orville Wright's diary entry that day, describing the flights, see ibid., pp. 394–97. For the Wright achievement, see Richard P. Hallion, "The Wright Brothers: How They Flew," *Invention and Technology* 19, no. 2 (Fall 2003): 18–37. For technical studies of the 1903 Flyer, see F.E.C. Culick and Henry R. Jex, "Aerodynamics, Stability, and Control of the 1903 Wright Flyer," and Frederick J. Hooven, "Longitudinal Dynamics of the Wright Brothers' Early Flyers: A Study in Computer Simulation of Flight," in Wolko, *The Wright Flyer*, pp. 19–77.

32. On the 1906 Wright patent in the United States, see Rodney K. Worrell, "The Wright Brothers Pioneer Patent," *American Bar Association Journal* 65 (October 1979): 1512–18. For patent filings abroad, see Crouch, *The Bishop's Boys*, p. 312.

33. For the difficulties of the Wright brothers in this period, see Crouch, *The Bishop's Boys*, pp. 301–26. See also Phaedra Hise, "The Wright Brothers: How They Failed," *Invention and Technology* 19, no. 2 (Fall 2003): 42–49. On European aviation at this time, see also Charles Gibbs-Smith, *The Rebirth of European Aviation, 1902–1908: A Study of the Wright Brothers' Influence* (London: Her Majesty's Stationery Office, 1974).

34. For Bell's research in aeronautics, see Alexander Graham Bell, "Aerial Locomotion," *National Geographic* 18, no. 1 (January 1907): 1–34. See also Robert V. Bruce, *Bell: Alexander Graham Bell and the Conquest of Solitude* (Ithaca: Cornell University Press, 1973), pp. 430–54.

35. On Curtiss, see C. R. Roseberry, *Glenn Curtiss: Pioneer of Flight* (Garden City, NY: Doubleday and Company, 1972), pp. 48–162.

36. For Wilbur Wright's public flights in 1908–9, and the 1908 tests by Orville Wright at Fort Myer, see Crouch, *The Bishop's Boys*, pp. 360–78, 406–8. On the formation of the Wright company and its training of military pilots, see ibid., pp. 395–410, 435–36.

37. For the invention of stick control and its superiority to the Wright lever system, see Malcolm J. Abzug and E. Eugene Larrabee, *Airplane Stability and Control: A History of the Technologies That Made Aviation Possible* (Cambridge: Cambridge University Press, 1997), pp. 5–6.

38. For the business and patent difficulties of the Wright brothers, see Crouch, *The Bishop's Boys*, pp. 411–67.

39. On the Curtiss-Wright dispute and Ford's involvement, see Roseberry, *Glenn Curtiss*, pp. 152–58, 257, 308–62; and Crouch, *The Bishop's Boys*, pp. 402–15, 461–62. For Orville's decision to sell the company, see Crouch, *The Bishop's Boys*, pp. 464–67. The company merged with the Curtiss firm in 1929.

40. On the flight of the rebuilt Langley plane and the Smithsonian's claim, see A. F. Zahm, "The First Man-Carrying Aeroplane Capable of Sustained Free Flight—Langley's Success as a Pioneer in Aviation," *Annual Report of the Board of Regents of the Smithsonian Institution . . . 1914* (Washington, DC: Smithsonian Institution, 1915), pp. 217–22. See also Tom Crouch, "The Feud between the Wright Brothers and the Smithsonian," *American Heritage of Invention and Technology* 2, no. 3 (Spring 1987): pp. 34–46.

41. Crouch, *The Bishop's Boys*, pp. 484–501, 526–29; and Jakab, *Visions of a Flying Machine*, pp. 220–21.

42. Anderson, *A History of Aerodynamics*, pp. 115 and 243. See also pp. 114–38, 192, 242–43.

CHAPTER SEVEN

Radio: From Hertz to Armstrong

1. For Maxwell's theory, see James Clerk Maxwell, *A Treatise on Electricity and Magnetism* (Oxford: Clarendon Press, 1873); and Paul J. Nahin, *The Science of Radio* (Woodbury, NY: American Institute of Physics Press, 1996), pp. 7–10.

2. For Hertz and his experimental work, see Heinrich Hertz, *Electric Waves* (1893; repr., New York: Dover Publications, 1962). For a description of Hertz's experiments, see Hugh G. J. Aitken, *Syntony and Spark: The Origins of Radio* (New York: John Wiley and Sons, 1976), pp. 48–79.

3. A standing wave forms when a transmitted wave and a reflected wave of the same amplitude and direction interact. For Hertz's measurements, see Hertz, *Electric Waves*, pp. 132–33.

4. The formula for frequency shown here assumes that inductance L and capacitance C are the only impedances to the current. The formula neglects the usually negligible resistance (in ohms) that causes heating. Hertz designed his receiving loop to have the same resonance as his dipole transmitting antenna. But he calculated the resonant frequency of the dipole antenna using a half-period $T = \pi\sqrt{PC/A}$ in which P is related to inductance, C is related to capacitance, and A is the speed of light. He got $T = \pi\sqrt{1902(15)}/3 \times 10^8 = 5.31/3 \times 10^8 = 1.77 \times 10^{-8}$ seconds, or a frequency of $f = 1/2\ T = 28.3 \times 10^6$ cycles per second or 28.3 MHz. Due to an error, C should have been $15/2 = 7.5$ cm and hence $f = 39.7$ MHz. See Hertz, *Electric Waves*, pp. 50–51 and 271. Later, on p. 133, he states that $T = 1.4 \times 10^{-8}$ seconds or $f = 35.6$ MHz. He does not give the calculation for this figure.

5. On Marconi and his early radio research, see G. Marconi, "Wireless Telegraphy," *Journal of the Institution of Electrical Engineers* 28 (1899): 273–97. See also Aitken, *Syntony and Spark*, pp. 179–297; and Sungook Hong, *Wireless: From Marconi's Black Box to the Audion* (Cambridge MA: MIT Press, 2001), pp. 17–23. For a biography of Marconi, see W. P. Jolly, *Marconi* (New York: Stein and Day, 1972). For spark transmission and reception, see Nahin, *Science of Radio*, pp. 24–37. Marconi's grounding of the transmitting antenna increased its capacitance.

6. On the dispute between Marconi and Lodge, see Hong, *Wireless*, pp. 25–51. Hong points out that before Marconi showed its usefulness for wireless telegraphy, some scientists tended to regard Hertzian waves in narrower terms as a substitute for optical signaling, for example, as a way to replace the light provided by coastal lighthouses.

7. On Marconi's subsequent career, see Jolly, *Marconi*, esp. p. 31. On his firm, see W. J. Baker, *A History of the Marconi Company* (New York: St. Martin's Press, 1972).

8. For Fessenden's life, see Frederick Seitz, *The Cosmic Inventor: Reginald Aubrey Fessenden (1866–1932)* (Philadelphia: American Philosophical Society, 1999).

9. On the Fessenden alternator, see James E. Brittain, *Alexanderson: Pioneer in American Electrical Engineering* (Baltimore: Johns Hopkins University Press, 1992), pp. 29–43.

10. On Fessenden's radio work, see R. A. Fessenden, "Wireless Telephony," *Proceedings of the American Institute of Electrical Engineers* 27 (1908): 1283–1358; and Hugh G. J. Aitken, *The Continuous Wave: Technology and American Radio, 1900–1932* (Princeton, NJ: Princeton University Press, 1985), pp. 40–86. On early crystal rectifiers, see Desmond P. C. Thackeray, "When Tubes Beat Crystals: Early Radio Detectors," *IEEE Spectrum* 20, no. 3 (March 1983): 64–69. Rectifiers

are also called "detectors" in radio. For an overview of radio engineering without calculus, see Abraham Marcus and William Marcus, *Elements of Radio* (New York: Prentice-Hall, 1943).

11. For Fessenden's Christmas 1906 radio broadcast, see Susan J. Douglas, *Inventing American Broadcasting, 1899–1922* (Baltimore: Johns Hopkins University Press, 1987), pp. 155–56. On amplitude modulation, see Marcus and Marcus, *Elements of Radio*, pp. 51–63, 605–18.

12. On the near sale of Fessenden's patents to AT&T, see Aitken, *The Continuous Wave*, pp. 76–79.

13. On Marconi's efforts to improve spark transmission, see Hong, *Wireless*, pp. 62–63, 90–107.

14. On tuning circuits and the reproduction of sound in radios, see Marcus and Marcus, *Elements of Radio*, pp. 31–42, 443–73.

15. On de Forest, see James A. Hijiya, *Lee de Forest and the Fatherhood of Radio* (Bethlehem, PA: Lehigh University Press, 1992); and Aitken, *The Continuous Wave*, pp. 162–249. On the collapse of de Forest's company, see ibid., pp. 185–94. Fessenden sued de Forest for using a liquid rectifier without permission. The company continued under de Forest's former partner under a new name. The firm collapsed after the former partner was convicted of mail fraud in 1910 and the American Marconi Company won a patent infringement suit a year later.

16. De Forest's audion patent was U.S. Patent No. 879,532. On the Edison effect, see J. B. Johnson, "Contribution of Thomas A. Edison to Thermionics," *American Journal of Physics* 28, no. 9 (December 1960): 763–73.

17. For the working of the Fleming diode, J. A. Fleming, *The Thermionic Valve and Its Developments in Radio-Telegraphy and Telephony*, 2nd ed. (New York: D. Van Nostrand, 1924), pp. 46–97. See also Marcus and Marcus, *Elements of Radio*, pp. 97–103.

18. For the working of the triode, see Lee de Forest, "The Audion: A New Receiver for Wireless Telegraphy," *Transactions of the American Institute of Electrical Engineers* 25 (1906): 735–79. See also Marcus and Marcus, *Elements of Radio*, pp. 105–19.

19. For an evaluation of de Forest's contribution to radio, see Robert A. Chipman, "DeForest and the Triode Detector," *Scientific American* 212, no. 3 (March 1965): 92–100.

20. On Armstrong, see Lawrence Lessing, *Man of High Fidelity* (Philadelphia: Lippincott, 1956). For his early life, see also Tom Lewis, *Empire of the Air: The Men Who Made Radio* (New York: Edward Burlingame Books, 1991), pp. 58–71.

21. The coiled section of the tuning circuit also had a greater number of turns that stepped up the voltage.

22. On the regenerative circuit, see E. H. Armstrong, "Some Recent Developments in the Audion Receiver," *Proceedings of the Institute of Radio Engineers* 3, no. 4 (September 1915): 215–46. See also Marcus and Marcus, *Elements of Radio*, pp. 121–29; and D. G. Tucker, "The History of

Positive Feedback: The Oscillating Audion, the Regenerative Receiver, and other applications up to around 1923," *Radio and Electronic Engineer* 42, no. 2 (February 1972): 69–80. Armstrong filed his patent on October 29, 1913. The patent, No. 1,113,149, was issued on October 6, 1914.

23. On the problem of howl in regenerative circuits and the solution of moving the plate coil, see Marcus and Marcus, *Elements of Radio*, pp. 123–25.

24. For Fessenden's work on heterodyning, see Aitken, *The Continuous Wave*, pp. 58–60.

25. On the superheterodyne receiver, see E. H. Armstrong, "A Study of Heterodyne Amplification for the Electron Relay," *Proceedings of the Institute of Radio Engineers* 5, no. 2 (April 1917): 145–59; and "The Super-Heterodyne: Its Origin, Development, and Some Recent Improvements," *Proceedings of the Institute of Radio Engineers* 12, no. 5 (October 1924): 539–52. See also Marcus and Marcus, *Elements of Radio*, pp. 231–47; and Paul J. Nahin, *The Science of Radio*, pp. 178–89. Armstrong arranged the heterodyning frequency to maintain a fixed distance from the frequency selected for reception, so that the difference frequency would remain the same. This enabled the radio to be designed to amplify a single difference frequency.

26. Armstrong filed his superheterodyne patent in the United States from France on February 8, 1919. The patent was issued on June 8, 1920, as No. 1,342,885. According to their entries in *The Dictionary of Scientific Biography* (New York: Scribner, 1970), de Forest won the Medal of Honor in 1915 and Armstrong in 1918.

27. On the formation of RCA, see Aitken, *The Continuous Wave*, pp. 281–431.

28. On Sarnoff's early life and career, see Lewis, *Empire of the Air*, pp. 89–117. See also Eugene Lyons, *David Sarnoff* (New York: Harper and Row, 1966); and Kenneth Bilby, *The General: David Sarnoff and the Rise of the Communications Industry* (New York: Harper and Row, 1986), pp. 9–67.

29. On Sarnoff's memorandum, see Lyons, *David Sarnoff*, pp. 91–92. For its implementation, see ibid., pp. 92–103; and Bilby, *The General*, pp. 49–52.

30. On the growth of radio broadcasting, see the papers in Lawrence W. Lichty and Malachi C. Topping, eds., *American Broadcasting: A Source Book on the History of Radio and Television* (New York: Hastings House, 1975). See also Douglas, *Inventing American Broadcasting, 1899–1922*; and Christopher H. Sterling and John Michael Kittross, *Stay Tuned: A Concise History of American Broadcasting*, 3rd ed. (Mahwah, NJ: Lawrence Erlbaum Associates, 2002). For early radio stations and frequencies, see Tom Kneitel, *Radio Station Treasury, 1900–1946* (Commack, NY: CRB Research, 1986). On improvements in radio technology during the 1920s, see David Rutland, *Behind the Front Panel: The Design and Development of 1920s Radios* (Philomath, OR: Wren Publishers, 1994).

31. On the shift from alternators to vacuum tubes for transmission, see Brittain, *Alexanderson*, pp. 176–78.

32. See Robert L. Hilliard, *The Federal Communications Commission: A Primer* (Boston: Focal Press, 1991), pp. 63–66. Station power determined the transmission range.

33. On the events leading up to the 1932 consent decree, see Aitken, *The Continuous Wave*, pp. 498–508.

34. On Armstrong's relationship to RCA in the early 1920s, see Lewis, *Empire of the Air*, pp. 165–67.

35. E. H. Armstrong, "Some Recent Developments in the Audion Receiver," *Proceedings of the Institute of Radio Engineers* 3, no. 4 (September 1915): 215–46.

36. On the Armstrong–de Forest patent dispute from 1915 to 1928, see Lewis, *Empire of the Air*, pp. 189–204. For the refusal of the Institute of Radio Engineers to accept the return of Armstrong's medal, see Tucker, "The History of Positive Feedback," p. 71.

37. For Sarnoff's position in the Armstrong–de Forest dispute, see Lewis, *Empire of the Air*, p. 204.

38. For the final 1934 Supreme Court decision, see ibid., pp. 204–14.

39. On the decision against Armstrong's claim to the superheterodyne receiver, see ibid., pp. 204–5.

40. On FM, see E. H. Armstrong, "A Method of Reducing Disturbances in Radio Signaling by a System of Frequency Modulation," *Proceedings of the Institute of Radio Engineers* 24, no. 5 (May 1936): 689–740. This paper is reprinted with other papers on FM in Jacob Klapper, ed., *Selected Papers on Frequency Modulation* (New York: Dover Publications, 1970). For Armstrong's development of FM, see Lewis, *Empire of the Air*, pp. 247–59.

41. For the conflict between Armstrong and Sarnoff over FM, see Lewis, *Empire of the Air*, pp. 260–68, 300–327. On Sarnoff's support for television, see Lyons, *David Sarnoff*, pp. 212–14.

42. On the postwar litigation leading to Armstrong's death, see Lewis, *Empire of the Air*, pp. 300–27.

CHAPTER EIGHT

Ammann and the George Washington Bridge

1. On Telford's bridges, see David P. Billington, *The Tower and the Bridge: The New Art of Structural Enginering* (New York: Basic Books, 1984), pp. 27–44. (Note: *The Tower and the*

Bridge is reprinted by Princeton University Press [1985], and page references to the original edition also apply to this paperback edition.)

2. On the Garabit Viaduct, see Elie Deydier, *Le Viaduc de Garabit* (Paris: Editions Gerbert, 1960). For the progression from Rouzat to the Eiffel Tower, see Gustave Eiffel, *La Tour Eiffel en 1900* (Paris: Masson et Cie, 1902), pp. 4–6.

3. On Roebling's Cincinnati Bridge (now the John A. Roebling Bridge), see John A. Roebling, *Report of John A. Roebling, Civil Engineer, to the President and Board of Directors of the Covington and Cincinnati Bridge Company* (Cincinnati, 1867).

4. For the Eads Bridge, see David P. Billington, *The Innovators: The Engineering Pioneers Who Made America Modern* (New York: John Wiley and Sons, 1996), pp. 144–54.

5. On the Brooklyn Bridge, see ibid., pp. 207–12; and David McCullough, *The Great Bridge* (New York: Simon and Schuster, 1972).

6. On the need for a Hudson River bridge, see O. H. Ammann, "General Conception and Development of Design," *Proceedings of the American Society of Civil Engineers* 59, no. 8, part 2 (October 1933): 16–21. This publication is subtitled "George Washington Bridge," *Transactions* 97 (1933). This journal issue is cited hereafter as *ASCE Transactions* 97 (1933).

7. On earlier plans to span the Hudson, see Ammann, "General Conception and Development of Design," pp. 2–9.

8. On Lindenthal and his earlier work, see Billington, *The Tower and the Bridge*, pp. 122–28.

9. For Lindenthal's design proposal, see Ammann, "General Conception and Development of Design," pp. 9–10. The cost of a tunnel is given on p. 11.

10. On Ammann's life and education, see David P. Billington, *The Art of Structural Design: A Swiss Legacy* (New Haven: Princeton Art Museum/Yale University Press, 2003), pp. 74–110. This chapter on Ammann is coauthored with Jameson W. Doig. On Ritter and his teaching, see ibid., pp. 16–29.

11. On Silzer, see Paul A. Stellhorn and Michael J. Birkner, eds., *The Governors of New Jersey, 1664–1974* (Trenton, NJ: New Jersey Historical Commission, 1982), pp. 194–96.

12. For Ammann's efforts to secure public and political support for the George Washington Bridge, see Jameson W. Doig and David P. Billington, "Ammann's First Bridge: A Study in Engineering, Politics, and Entrepreneurial Behavior," *Technology and Culture* 35, no. 3 (July 1994): 537–70. For a more general discussion of public entrepreneurship, see Jameson W. Doig

and Erwin C. Hargrove, eds., *Leadership and Innovation* (Baltimore: Johns Hopkins University Press, 1987).

13. For an account of the early Port Authority, see Erwin W. Bard, *The Port of New York Authority* (New York: Columbia University Press, 1942). Bard does not describe Ammann's role prior to his selection as designer for the bridge. The best and most recent study of the Port Authority is Jameson W. Doig, *Empire on the Hudson: Entrepreneurial Vision and Political Power at the Port of New York Authority* (New York: Columbia University Press, 2001). Doig highlights Ammann's pivotal role in this history. The Port Authority became the Port Authority of New York and New Jersey in 1972.

14. Doig and Billington, "Ammann's First Bridge," pp. 563–65.

15. Ammann, "General Conception and Development of Design," p. 10. References in the article to the War Department are to the U.S. Army Corps of Engineers.

16. See ibid., pp. 1–65. More detailed description of Ammann's design may be found in the articles that follow in the *ASCE Transactions* 97 (1933).

17. J.A.L. Waddell, *Bridge Engineering*, 2 vols. (New York: John Wiley and Sons, 1916), 1:117.

18. See Leon S. Moissieff, "The Towers, Cables, and Stiffening Trusses of the Bridge over the Delaware River between Philadelphia and Camden," *Journal of the Franklin Institute* 200, no. 4 (October 1925): 436–66. Ammann was probably aware of this study a year before its publication.

19. For Ammann's traffic load analysis, see Allston Dana, Aksel Andersen, and George M. Rapp, "George Washington Bridge: Design of Superstructure," *ASCE Transactions* 97 (1933): 103–4.

20. See *Standard Specifications for Highway Bridges* (Washington, DC: American Association of State Highway Officials, 1931), p. 176.

21. Doig and Billington, "Ammann's First Bridge," pp. 553–54.

22. For the cable strength, see Dana, Andersen, and Rapp, "George Washington Bridge: Design of Superstructure," p. 109.

23. The deflection theory was formulated by Wilhelm Ritter in Switzerland and Josef Melan in Austria during the 1880s. A presentation for American engineers appeared in J. B. Johnson, C. W. Bryan, and F. E. Turneaure, *The Theory and Practice of Modern Framed Structures*, 9th ed. (New York: John Wiley and Sons, 1911), part 2, pp. 276–321. For Ammann's own embrace of it, see Ammann, "General Conception and Development of Design," pp. 42–43; and Doig and Billington, "Ammann's First Bridge," pp. 558–59. The table below (ibid., p. 559), gives the bending moment in the stiffening trusses at midspan and shows the effects of thinness:

Kip Feet	Assumptions
2,390,400	Truss with no cable support
56,150	Truss with movable cable support (deflection theory)
6,980	More flexible truss with movable cable support (deflection theory)

24. Ammann, "General Conception and Development of Design," pp. 61–62.

25. For the influence of deflection theory on later bridges, see S. G. Buonopane and D. P. Billington, "Theory and History in Suspension Bridge Design from 1823 to 1940," *Journal of Structural Engineering* 119, no. 3 (March 1993): pp. 954–77.

26. The following table shows the trend toward lighter and thinner decks:

Bridge	Date Built	Width (ft.)	Span L (ft.)	Depth h (ft.)	Ratio of h/L
Williamsburg	1903	118	1,600	40.0	1/40
Manhattan	1909	120	1,470	24.5	1/60
Bear Mountain	1924	55	1,630	25.9	1/63
Delaware River	1926	89	1,750	27.8	1/63
Ambassador	1929	60	1,850	22.0	1/84
George Washington	1931	106	3,500	10.0	1/350
Golden Gate	1937	90	4,200	25.0	1/168
Deer Island	1939	25	1,080	6.5	1/166
Bronx-Whitestone	1939	75	2,300	10.9	1/210
Tacoma-Narrows	1940	40	2,800	8.0	1/350

Source: John Paul Hartman, "History and Esthetics in Suspension Bridges," *Journal of the Structural Division, Proceedings of the ASCE,* 104, 7, Proc. Paper 13857 (March 1979): 1174–76. Depth *h* refers to the vertical depth of the deck.

27. For the Tacoma Narrows collapse, see F. B. Farquharson, "Aerodynamic Stability of Suspension Bridges with Special Reference to the Tacoma Narrows Bridge," *University of Washington Engineering Bulletin Experiment Station Bulletin,* 116, part 1, investigations prior to October 1941. No. 116 appeared in four more parts published in 1950, 1952, 1954, and 1955. Aware that the bridge was in danger, Professor Farquharson was at the bridge at the time of its failure.

28. For Ammann's Verrazano Narrows Bridge, see David P. Billington, *The Tower and the Bridge,* pp. 137–38. Decks often consist now of a "hollow box" in which the roadway and a parallel surface under the deck are connected by sides that are tapered out like airplane wings. The box construction lends both torsional and longitudinal stiffness, and the tapered sides deflect wind much more efficiently than flat vertical sides.

29. O. H. Ammann, "George Washington Bridge: General Conception and Development of Design," pp. 38–39.

30. On the need for historical awareness, see David P. Billington, "History and Esthetics in Suspension Bridges," *Journal of the Structural Division, Proceedings of the American Society of Civil Engineers*, 103, ST8, Proc. Paper 13143 (August 1977): 1655–72. This paper provoked a discussion in subsequent numbers that concluded with David P. Billington, "History and Aesthetics in Suspension Bridges: Closure," 105, ST3, Proc. Paper 14404 (March 1979): 671–87.

31. From the horizontal force, $H = qL^2/8d$, the tension in the cable may be computed and then divided by the allowable stress to determine the cross-sectional area of the cable (sidebar 8.4). A higher cable sag d will permit a smaller cross-sectional area A in the cable and will reduce the weight needed by the anchorages. However, raising the height of the towers will increase the amount of steel needed in them.

32. For the Swiss tradition of which Ammann was an exemplar, along with Robert Maillart, Heinz Isler, and Christian Menn, see again Billington, *The Art of Structural Design: A Swiss Legacy*, with chapter 3 written with Jameson W. Doig.

CHAPTER NINE

Eastwood, Tedesko, and Reinforced Concrete

1. On Hoover Dam, see Joseph E. Stevens, *Hoover Dam: An American Adventure* (Norman: University of Oklahoma Press, 1988). The dam was known as Boulder Dam from 1933 until 1947.

2. For the origins of reinforced concrete, see David P. Billington, *The Tower and the Bridge: The New Art of Structural Engineering* (New York: Basic Books, 1984), pp. 148–51.

3. Max Bill, *Robert Maillart* (Erlenbach-Zürich: Verlag für Architektur AG/Les Editions d'Architecture SA, 1949), p. 26.

4. For Maillart and his work, see David P. Billington, *Robert Maillart: Designer, Builder, Artist* (New York: Cambridge University Press, 1997).

5. David P. Billington, "Maillart and the Salginatobel Bridge, Switzerland," *Structural Engineering International* 1, no. 4 (November 1991): 46–50.

6. On Eastwood and his work, see Donald C. Jackson, *Building the Ultimate Dam: John S. Eastwood and the Control of Water in the West* (Lawrence: University Press of Kansas, 1995).

7. On nineteenth-century dam design, see Norman Smith, *A History of Dams* (Secaucus, NJ: Citadel Press, 1972). For the major types of dams, see Donald C. Jackson, *Great American Bridges and Dams* (New York: John Wiley and Sons, 1988), pp. 41–53.

8. On Huntington, see William B. Friedricks, *Henry E. Huntington and the Creation of Southern California* (Columbus: Ohio State University Press, 1992).

9. Jackson, *Building the Ultimate Dam*, pp. 66–83. The savings in materials of Eastwood's design are given on pp. 71–72.

10. On Hume Lake, see ibid., pp. 85–98.

11. On Big Bear, see ibid., pp. 98–104.

12. For a report of the Eastwood design, see "The Big Meadows Dam," *Journal of Electricity, Power, and Gas* 27 (September 30, 1911): 287–89. See also Jackson, *Building the Ultimate Dam*, pp. 109–33, and on Schuyler and Noble, pp. 114–15. The former had published a standard book on reservoirs and dams: James D. Schuyler, *Reservoirs for Irrigation, Water-Power, and Domestic Water-Supply* (New York: John Wiley and Sons, 1901).

13. Jackson, *Building the Ultimate Dam*, p. 123.

14. Ibid., p. 127.

15. On the Mountain Dell competition, see ibid., pp. 146–51. Bids are round numbers. Eastwood began to describe his innovation in the engineering press.

16. On the Lake Hodges Dam, see Jackson, *Building the Ultimate Dam*, pp. 160–67.

17. On the Littlerock Dam, see ibid., pp. 197–208. Eastwood replaced his radial design with a straight crested dam.

18. On Webber Creek, see ibid., pp. 219–23.

19. On the St. Francis Dam collapse, see Charles F. Outland, *Man-Made Disaster: The Story of the St. Francis Dam* (Glendale, CA: Arthur H. Clark, 1963).

20. Jackson, *Building the Ultimate Dam*, pp. 241–44. See also Fred D. Pyle, "Hodges Dam Strengthened," *Engineering News-Record* 117 (November 5, 1936): 644–47.

21. David P. Billington, "History and Aesthetics in Suspension Bridges," *Journal of the Structural Division* (American Society of Civil Engineers) 103, no. ST8 (August 1977): 1–18.

22. Lars Jorgensen, "The Record of 100 Dam Failures," *Journal of Electricity* 44, no. 6 (March 15 and April 1, 1920): 274–76, 320, and 321.

23. At the time, Dyckerhoff & Widmann employed Franz Dischinger (1887–1953), Ulrich Finsterwalder (1897–1988), Wilhelm Flügge (1904–90), and Hubert Rüsch (1903–79). The first two were designer-builders while the last two became academics, Flügge writing a pioneering text on thin shells and Rüsch doing advanced research on reinforced-concrete structures. See D. P. Billington, "Anton Tedesko: Thin Shells and Esthetics," *Journal of the Structural Division* (American Society of Civil Engineers) 108, ST11 (November 1982): 2541–44. The original stimulus to the firm's work in thin concrete shell design was a 1922 scale model dome in

Jena, Germany. The Zeiss Optical Company commissioned the dome to test the functioning of a new planetarium to be placed in the German Museum in Munich. Dyckerhoff & Widmann built the dome and then worked out a mathematical theory for such shells. The theory and construction technique came to be known as the "Zeiss-Dywidag System." See *Shell-Vaults, System "Zeiss-Dywidag,"* (Wiesbaden-Berlin: Dyckerhoff & Widmann A.G., 1931), a sixty-page promotional brochure, Anton Tedesko Papers, Princeton Maillart Archive, Princeton University, Princeton, New Jersey.

24. Anton Tedesko, "A Chronicle, Part III," 1984, pp. 1–38. Unpublished manuscript in the Tedesko Papers.

25. D. P. Billington, "Anton Tedesko: Thin Shells and Esthetics."

26. Anton Tedesko, "A Chronicle, Part IV," 1986, pp. 2–3. Unpublished manuscript in the Tedesko Papers.

27. Starline advertisement flyers, Harvard, Illinois, 1934, in the Tedesko Papers.

28. "Reinforced Concrete Shell Roof over Unobstructed Dairy Floor," *Concrete* 42, no. 7 (July 1934): 3–4. For the tests, see "Thin Concrete Shell Roof Tested under Large Unsymmetrical Load," *Engineering News-Record* (November 7, 1935): 635. See also A. Tedesko, "Memorandum on the Construction Century of Progress Barn Final Report," May 10, 1934, 3 pages, Tedesko Papers.

29. R. L. Bertin, "Construction Features of the Zeiss Dywidag Dome for the Hayden Planetarium Building," *Journal of the American Concrete Institute* 31 (May–June 1935): 449–60.

30. On Milton Hershey, see James D. McMahon, *Built on Chocolate: The Story of the Hershey Chocolate Company* (Los Angeles: General Publishing Group, 1998).

31. By using chocolate workers, the greatest part of the arena's cost, construction labor, was donated by the client, so the cost of the Hershey Arena is difficult to compare with other projects. Roberts and Schaefer agreed to do design and construction supervision for 14 ¢ per square foot of covered area or $75,000 \times 0.14 = \$10,500$. Later the area increased to 84,933.33 square feet, so their fee increased accordingly. See Tedesko to Hershey Lumber Products, January 21, 1936; Roberts to Hershey Lumber Products, January 25, 1936; and Tedesko to Hershey Lumber Products, February 3, 1936, Tedesko Papers.

32. On the design and building of the Hershey Arena, see Tedesko, "A Chronicle, Part IV," pp. 25–29. See also D. Paul Witmer "Sports Palace for Chocolate Town," and Anton Tedesko, "Z-D Shell Roof at Hershey," *Architectural Concrete* (Portland Cement Association) 3, no. 1 (1937): 1–11; "Thin-Shell Barrel Roof," *Construction* (April 1937): 44–47; and Anton Tedesko, "Large Concrete Shell Roof Covers Ice Arena," *Engineering News-Record*, April 8, 1937.

33. In February 1981, Tedesko sent these 1936 calculations to Hershey and a copy is deposited in the Tedesko Papers at the Princeton Maillart Archive. His calculations of the horizontal forces produced slightly different numbers than those in sidebars 9.3 and 9.4 because his analysis was more accurate than the one presented here, but the difference is less than 1 percent of the values of H given in the figures here. For a full discussion of the history and calculations of the arena, see Edmond P. Saliklis and David P. Billington, "Hershey Arena: Anton Tedesko's Pioneering Form," *Journal of Structural Engineering* 129, no. 3 (March 2003): 278–85.

34. Tedesko, "A Chronicle, Part IV."

35. For Tedesko's recollections of the Hershey Arena, quoted here, see ibid., pp. 25–29.

36. On Tedesko's later career, see Billington, "Anton Tedesko: Thin Sheels and Esthetics," pp. 2541–44. Tedesko also built hangers for the U.S. Army Air Force at bases around the United States during World War II.

37. Tedesko, "A Chronicle, Part IV."

38. Ibid., p. 34.

39. Ibid., p. 29. See also Anton Tedesko, "Thin Concrete Shell Roof for Ice Skating Arena," *Engineering News-Record*, February 16, 1939. The senior author, a native of a town adjoining Ardmore, met Schwertner in 1952 through a Princeton classmate, Bill Borden. The senior author then met Tedesko, worked for him from 1952 until 1960, and remained a close friend until his death in 1994.

40. Eric Molke, "Elliptical Concrete Domes for Sewage Filters," *Engineering News-Record*, November 9, 1939, pp. 623–25. See also Anton Tedesko, "Point-Supported Dome of Thin Shell Type," *Engineering News-Record*, December 7, 1939.

41. Anton Tedesko, "Tire Factory at Natchez," *Engineering News-Record*, October 26, 1939. The compression stress given was 282 psi and the concrete strength at decentering was 2,500 psi. After a few weeks the strength would be at least 3,000 psi.

CHAPTER TEN

Streamlining: Chrysler and Douglas

1. For the aerodynamics of the 1930s, see Walter Stuart Diehl, *Engineering Aerodynamics* (New York: Ronald Press Company, 1940). On drag, see John D. Anderson Jr., *A History of*

Aerodynamics and Its Impact on Flying Machines (Cambridge: Cambridge University Press, 1997), pp. 73–74, 320–21; and on streamlining, pp. 321–28.

2. For Walter Chrysler, see Vincent Curcio, *Chrysler: The Life and Times of an Automotive Genius* (New York: Oxford University Press, 2000), pp. 1–124, 215–41.

3. Ibid., pp. 241–335.

4. On Zeder, Skelton, and Breer, see Richard P. Scharchburg, "Zeder-Skelton-Breer Engineering," in George S. May, ed., *Encyclopedia of American Business History and Biography: The Automobile Industry, 1920–1980* (New York: Facts on File, 1989), pp. 500–508; and Carl Breer, *The Birth of the Chrysler Corporation and Its Engineering Legacy*, ed. Anthony J. Yanik (Warrendale, PA: Society of Automotive Engineers, 1995).

5. On the hiring of Zeder, Skelton, and Breer, see Richard P. Scharchburg, "Zeder-Skelton-Breer Engineering," in May, *Encyclopedia of American Business History and Biography: The Automobile Industry, 1920–1980*, p. 504. For the Chrysler Six, see Breer, *The Birth of the Chrysler Corporation*, pp. 79–90; and Curcio, *Chrysler*, pp. 297–313. Richard P. Scharchburg, "Walter Percy Chrysler," in May, *Encyclopedia of American Business History and Biography: The Automobile Industry, 1920–1980*, gives prices and sales figures on pp. 61–62. The car competed primarily against Buicks.

6. For Chrysler's rise to Big Three status in 1927–28, see Curcio, *Chrysler*, pp. 361–99. The Plymouth sold for between $670 and $725.

7. For the innovations at Chrysler, see Breer, *The Birth of the Chrysler Corporation*, pp. 91–142. Apart from the Airflow, these were incremental in nature.

8. On the Airflow car, see Howard S. Irwin, "History of the Airflow Car," *Scientific American* 237, no. 2 (August 1977): 98–105; Breer, *The Birth of the Chrysler Corporation*, pp. 143–75; and James J. Flink, "The Path of Least Resistance," *Invention and Technology* 5, no. 2 (Fall 1989): 34–44. We have assumed a coefficient of 0.015 for a paved road and estimated the frontal area of the Chrysler Airflow to be twenty-eight square feet.

9. On the failure of the Airflow, see Curcio, *Chrysler*, pp. 542–57.

10. For Douglas, see Frank Cunningham, *Sky Master: The Story of Donald Douglas* (Philadelphia: Dorrance, 1943); and Wilbur M. Morrison, *Donald W. Douglas: A Heart with Wings* (Ames: Iowa State University Press, 1991). For Orville Wright's second demonstration at Ft. Myer on July 30, 1909, and the job offer from Edison, see ibid., pp. 4–6.

11. On the 1924 World Cruiser, see Morrison, *Donald W. Douglas*, pp. 30–33. For the specifications of the World Cruiser, see C. G. Grey, ed., *[Jane's] All the World's Aircraft* (London: Sampson Low, 1925), pp. 249–50.

12. On the Ford Trimotor, see Allan Nevins and Ernest Frank Hill, *Ford: Expansion and Challenge, 1915–1933* (New York: Charles Scribner's Sons, 1957), pp. 238–47; and C. G. Grey and Leonard Bridgman eds. *Jane's All the World's Aircraft, 1930*, pp. 281–83.

13. See Louis Breguet, "Aerodynamical Efficiency and the Reduction of Air Transport Costs," *Aeronautical Journal* 26 (1922): 307–13; and B. Melvill Jones, "The Streamline Airplane," *Aeronautical Journal* 32 (1929): 358–85.

14. On Boeing, see Robert Redding and Bill Yenne, *Boeing: Planemaker to the World* (London: Arms and Armour Press, 1983). For the Monomail airplane, see C. G. Grey and Leonard Bridgman, eds. *Jane's All the World's Aircraft, 1931*, pp. 252–53. For the Boeing 247, see *Jane's All the World's Aircraft, 1934*, pp. 259–60 (for the 247-D, which had an improved propeller); and R. E. G. Davies, *Airlines of the United States since 1914* (Washington, DC: Smithsonian Institution Press, 1972), pp. 180–83.

15. On the death of Rockne, see ibid., pp. 93–94; and for the TWA negotiations with Douglas, see ibid., pp. 183–84. See also Frederick Allen, "The Letter That Changed the Way We Fly," *Invention and Technology* 4, no. 2 (Fall 1988): 6–13. For the TWA specifications, see Peter M. Bowers, *The DC-3: 50 Years of Legendary Flight* (Blue Ridge Summit, PA: Aero/TAB Books, 1986), p. 22.

16. For the role of NACA in aviation research, see Anderson, *A History of Aerodynamics*, pp. 294–96, 301–4, 328–69. For the use of NACA research by Douglas in the DC series planes, see ibid., p. 358.

17. On the Douglas Transport and DC-1, see Grey and Bridgman, *Jane's All the World's Aircraft, 1934*, pp. 278–79. For its test flight, see Cunningham, *Sky Master*, pp. 220–21.

18. On the DC-2, see Wilbur M. Morrison, *Donald W. Douglas*, pp. 77–90; Bill Yenne, *McDonnell Douglas: A Tale of Two Giants* (New York: Crescent Books, 1985), pp. 84–91; and C. G. Grey and Leonard Bridgman, eds., *Jane's All the World's Aircraft, 1935*, pp. 300–301. For a comparison of the DC-1 and Boeing 247, see again Bowers, *The DC-3*, p. 22.

19. On the DST and DC-3, see Yenne, *McDonnell Douglas*, pp. 92–118; and C. G. Grey and Leonard Bridgman, eds., *Jane's All the World's Aircraft, 1937* pp. 292–94. For the figure on planes in service in 1942, see Davies, *Airlines of the United States since 1914*, p. 608.

20. For the technical data on the Wright Flyer, see chapter 6. For the DC-3, see Grey and Bridgman, *Jane's All the World's Aircraft, 1937*, pp. 292–94.

21. See Donald J. Bush, *The Streamlined Decade* (New York: George Braziller, 1975). On the New York World's Fair, see Helen A. Harrison, guest curator, *Dawn of a New Day: The New York World's Fair, 1939/40* (New York: Queens Museum/New York University Press, 1980).

Index

References to sidebars' pages are listed in bold type

Ader, Clement, 238n3

aerodromes, 106–7, 120, 124. *See* Langley, Samuel P.

aerodynamics: in airplanes, 103–4, **105**, 110, 114, **115**, 116, 119, **122**, 127–28, 208–9; in automobiles, **93**, 203, **204**; in bridges, 169, 172, 249n28. *See also* streamlining

Airflow car. *See* Chrysler Airflow

airplane. *See* aviation

Alexanderson, Ernst, 135

alkylation, 78. *See also* oil refining

alternating current (A.C.), 11, 13, 14, **16**, 25–34. *See also* electric circuits; electric power; Steinmetz, Charles; Tesla, Nikola; Westinghouse, George

alternators, **16**, 135

American Chemical Society, 71

American Telephone and Telegraph Company (AT&T), 33, 51, 54; and long-distance telephony, 51–53; and radio, 53, 141, 143; regulation of, 54. *See also* Bell Telephone Company

Ammann, Othmar, 11, 160, 161; and campaign for George Washington Bridge, 161–63; and design of George Washington Bridge, **159**, 163–65, **166**, **167**, **168**, 169, 170–71; later bridges of, 169, 172–75, 249n28. *See also* George Washington Bridge

Ampère, André-Marie, 14

amplitude modulation (AM), 135–36, **137**. See also Fessenden, Reginald Aubrey; radio; radio transmission

Anderson, John D. Jr., 127–28

angle of attack, 103, **105**. *See also* aviation, principles of

antenna coupler, **139**, 143. *See also* radio reception

arc lighting, 14, 17, 19, **20**, 22, 224n7, 225n10. *See also* electric lighting

Armstrong, Edwin Howard, 12, 143, 145, 152, 153–54; and frequency modulation (FM), 153; patent litigation of, 149–51, 153–54; and the regenerative circuit, 143, **144**, 147, 150; and the superheterodyne receiver, **144**, 145–46, 147, 151, 245n25. *See also* de Forest, Lee; Sarnoff, David

Ardmore, Pennsylvania, thin-shell roof, 197

assembly-line manufacturing. *See* Ford, Model T: manufacturing of,

AT&T. *See* American Telephone and Telegraph Company (AT&T)

audion. *See* de Forest, Lee; triode

automobile (gasoline-powered), xv, 6, 78, 79; early cars, 65, 80, 82–83, 86–87; engine, four-stroke cycle in, 79, **81**; Ford Model T, 65, 79, 87–88, 89, **90**, **91**, 92, **93**, **94**, 95, 100; fuel needs of, 57, 65, 75; later innovations in, 83,

thermal cracking research, 65–66, 68–69. *See also* Burton process; oil refining: thermal cracking; Standard Oil Company; Standard Oil of Indiana

Burton process, 57; control-volume analysis, as example of, 71–72; development of, 65–69, **70**, 72, 78; limitation of, as batch process, 72; stills in operation, 73

camber (in airplane wings), 110, 112

Campbell, George A., 52; and inductive loading, 52–53, 54, 56, 230n26; and wave filter, 53

capacitance, 52; in Hertz experiment, 131–32, 243n4; in radio tuning, 138, **139**. *See also* electric circuits

Cape Canaveral, Vertical Assembly Building at, 198

car. *See* automobile

carbon telephone transmitter, 7, 48. *See also* Edison, Thomas Alva

Cardozo, Benjamin, 151

Carnegie, Andrew, 8, 57

carrier wave, 135–36, **137**. *See also* radio, reception

catalytic cracking, 75–76, **77**, 78, 233n34; fixed-bed, 78; moving-bed, 78

Cayley, Sir George, 103–4, 110

Central Pacific Railroad, 178

Century of Progress World's Fair (Chicago, 1933–34), 187

Chanute, Octave, 108, 113, 118

Chevrolet, Louis, 99

Chrysler, Walter P., 201–2; and Chrysler Motors, 201–3; and Zeder-Skelton-Breer group, 201–2

Chrysler Airflow, 199, 201, 205; aerodynamic design of, 203, **204**, 205; commercial failure of, 205

Chrysler Building, 217, 219

Chrysler Corporation, 11–12, 201–3

Chrysler Six, 202. *See also* Chrysler Corporation

Cincinnati Bridge (now the John A. Roebling Bridge), 156

circuits. *See* electric circuits

Clermont, 5

coal gas: in lighting, 13, 14, 58; in Otto engine, 6, 79, **81**

coherer, 132. *See also* radio reception

Columbia Broadcasting System (CBS), 147

Columbian Exposition (Chicago, 1893), 30

concrete, 177. *See also* reinforced concrete

control-volume analysis, 71–72

Coolbaugh, John and Kenneth, 149

Coolidge, William D., 33

Corliss, George, 4

Corliss engine, 4, 5

Cornell, Ezra, 38, 228n6

Couzens, James, 85–86

Craigellachie Bridge, 156

Curtiss, Glenn, 124, 126–27

Daimler, Gottlieb, 80

dams: arch, 178, **180**; earth, 178, **180**; "flat slab," 178; gravity, 178, **180**; multiple-arch, 178, 181–82, **183**, 185–86. *See also* Eastwood, John; Freeman, John. *See also specifically listed individual dams*

Davy, Sir Humphrey, 14

deflection theory, 165, **168**, 169, 248n23, 249n26. *See also* George Washington Bridge; Tacoma Narrows Bridge

de Forest, Lee, 53, 138, 140, 143; patent conflicts of, with Armstrong, 150–51; —, with Fessenden, 140, 244n15; triode (audion), develop-

"Edison Effect," 140–41

Edison General Electric Company, 25, 30. *See* General Electric Company; Westinghouse Company

Edwards, Nelson, 197

Eiffel, Gustave, 128, 156

electric cars, 82, 83

electric circuits, xix, 14, **15**; alternating current (A.C.), 11, 13, 14, **16**, 25–28, **29**, 29–31, 131, 226n21; in automobiles, **90**; capacitance and inductance, 52, 131–32, 138, **139**, 243n4; direct current (D.C.), 11, 13, 14, **16**, 17, 25, 28, 31, 131; parallel circuit, 17, 19, **20**, 22, 220–22; in radio, 129, **130**, 131, 132, **139**, **142**, 143, **144**; resistance, 14, **15**, 19, **21**, 22; resonance, 138, **139**; series circuit, 17, **20**; in telegraphy, 37–38, **39**, **41**; in telephony, 43, 48, **49**; in transformers, 28–29. *See also* electric lighting; electric power; electromagnetism

electric lighting, xv, 6, 13, 14; arc lighting, 14, 17, **20**; incandescent lighting, 14, 17, 19, **20**, 22, **24**, 25, 30; high-resistance filament in, 19, **21**, 22; high-vacuum bulb in, 22; later improvements to, 33. *See also* arc lighting; Edison, Thomas Alva; incandescent lighting; Swan, Sir Joseph; Westinghouse, George

electric power, xv, 13; alternators, **16**, 135; dynamos, **16**, 17, 23, **24**, 25; generation of, 14, **16**, 17, **24**, 25; Edison (direct-current) system, 23, **24**, 25–26, 29–30; transformers and, 28, **29**; Westinghouse (alternating-current) system, 25–30.

electric (electric-arc) welding, 72

electromagnetism, 14, **15**; in generating electricity, 14, **16**, 17, 28; Henry and, 37, **39**; in telegraphy, 37–38, **39**, 40, **41**; in telephony, 43, 48, **49**, 52–53; in radio waves, 129, **130**, 131–32

engine. *See* internal-combustion engine; steam engine

engine knock, 73

engineering: as design, xvii; as four great ideas (structures, machines, networks, processes), xvi, 8–10; as narrative of great works, xviii; as normal and radical design, xvii. *See also* innovation; science

English Visible Speech (Bell), 36

Ethyl Corporation, 74

Euler, Leonhard, 127

Evans, Harold, xvi

ExxonMobil, xv. *See* Standard Oil of New Jersey

Faraday, Michael, 14, 131

Farquharson, F. B., 249n27

Federal Communications Commission, 148

Federal Radio Commission, 148

Federal Trade Commission, 149

Fessenden, Reginald Aubrey, 135, 136, 138, 143; and amplitude modulation (AM), 136; and broadcasting, 136, 147; and carrier wave, recognition of need for, 135; and heterodyning, 145–46; and high-frequency alternator, 135; patent conflict with de Forest, 140, 244n15

filament. *See* electric lighting

Finsterwalder, Ulrich, 251n23

first law of thermodynamics, 68

Fleming, John Ambrose, 140–41, **142**

Flügge, Wilhelm, 251n23

Ford, Henry, 11, 65, 79, 83, 85; and aviation, 208; early companies and car models, 84–87; and Ford Model T, 87–88, 92, 95; and labor, 95, 96–97, 236n21; and moving assembly line, 92, 95; "quadricycle" and racecars, 84; rigidity and anti-Semitism in later years, 11–12, 100; rivalry

261

169, 173; deflection theory, influence on design of, 165, **168**, 169; design of, **159**, 163–65, **166**, **167**, 169, 170, 171, 173–74; need for, 158, 160–61; public campaign for, by Ammann, 162–63; steel calculations for, 165, **167**; traffic load estimate for, 164–65, **166**, 173. *See also* Ammann, Othmar

Gibbs, Josiah Willard, 138; Josiah Willard Gibbs Medal, 71

Gibson, James, 188

GM. *See* General Motors Corporation (GM)

Golden Gate Bridge, 176, 186

Gould, Jay, 50

Gramme, Zénobe Théophile, 17

Grant, Ulysses S., 4

Gray, Elisha, 47, 54; patent conflict with Bell, 47–48, 50, 229n17; telephone of, 47–48, **49**. *See also* Bell, Alexander Graham; Western Union

"Great Aerodrome" (of Langley), 120, 124

Great Western Power Company, 181

Grove, Sir William, 14

Harris, King and Lawrence, 55

Harlem Board of Commerce, 163

harmonic telegraph: Bell's, 40, **41**, 42–43; Gray's, 47. *See also* Bell, Alexander Graham, telegraph research of; Gray, Elisha

Harvard University, 37

Hayden Planetarium, 187–88, 189

Heaviside, Oliver, 53

Hell Gate Bridge, 158, 160, 161

Helmholtz, Hermann von, 36

Henry, Joseph, 131; and Bell, Alexander Graham, 43; and electromagnetic telegraph, 37, **39**. *See also* electromagnetism

Hershey, Milton, 188–89

Hershey Arena, 188–90; construction of, 190–91, 193–96; costs of, 252n31; design of, 190; labor, Hershey chocolate workers used for, 189–90, 193; forces and stresses in, **191**, **192**, 253n33. *See also* Tedesko, Anton

Hertz, Heinrich, 129, **130**, 131–32, 154

heterodyning (in radio), **144**, 145–46, 245n25. *See also* Armstrong, Edwin Howard; Fessenden, Reginald Aubrey

Hibbing, Minnesota, thin-shell domes, 197

Higgs, Paget, 220–22

Holland, Clifford, 160

Holland Tunnel, 160

Hong, Sungook, 243n6

Hooper, Stanford, 146

horizontal force (in structure), 10–11, 156, **157**; in George Washington Bridge, **167**, 250n31; in Hershey Arena, 191, 253n33

horsepower (hp), 10-11, 88, 92, 120; in Chrysler Airflow, **204**; in Douglas DC-3, **214**, **215**; in Ford Model T, 88, **91**, 92, **93**; formulas for: brake hp, 88, 92, 236n15; —: indicated hp, 10, 11, 88, **91**, 235n14; —: thrust hp (airplanes), 117–18, **119**, **122**, **214**, **215**; —: traction hp (cars), 92, **93**, **204**, 236n16; in Wright Flyer, 117–18, **119**, **122**, **215**

Houdry, Eugene, 57, 72, 75–76, 78

Houdry process, 75–76, **77**, 78. *See also* Houdry, Eugene; oil refining

Hubbard, Gardiner, 40, 42, 45, 50

Hubbard, Mabel, 40, 42, 45, 50; marriage to Alexander Graham Bell, 45

Huber, Walter, 186

Hudson, Henry, 124

Hume Lake Dam, 181, 182, **183**

Radio Corporation of America (RCA), 146–49;
    Sarnoff and, 146–51, 153, and patent conflicts
    with E. H. Armstrong, 149–51, 153–54
radio frequency spectrum, 133
railroads, 5–6, 61, 156, 158; external-combustion
    engines used with, 6, **81**
reactances. *See* electric circuits
refining of oil. *See* oil refining
reforming. *See* oil refining
regenerative circuit, 143, **144**, 244n21
reinforced concrete, xv, 175, 176; in dams,
    181–82, **183**, 184–86; mass versus form in,
    177–78; in thin shells, 186–191, **191**, **192**,
    193–98
Reis, Philip, 43
resonance (in radio), 138, **139**; in Hertz experi-
    ments, 131–32
Righi, Augusto, 132
Ritter, Wilhelm, 160
Roberts and Schaefer Company, 187, 188, 197
Rockefeller, John D., 6, 57, 62; consolidation of
    refining industry by, 61, 63; controversy about,
    63; innovation and, 64
Rockne, Knute, 210
Roebling, John A., 156, 173
Rogers, F. M., 65
rolling resistance, 92, **93**
Roosevelt, Eleanor, 196
Roosevelt, Franklin, 196
Roosevelt, Theodore, 63
Rüsch, Hubert, 251n23

St. Francis Dam, 186
Salginatobel Bridge, 178
San Joaquin Electric Company, 178
Sanders, George, 37

Sanders, Thomas, 37, 42, 45
Sarnoff, David, 146–47, 148, 151; broadcasting,
    vision of, 147; conflicts with Armstrong,
    149–51, 153–54; television and, 153
Schuyler, James D., 182
Schwertner, Charles, 197, 253n39
science: as discovery, xvii; and engineering educa-
    tion, xviii; contributions following radical inno-
    vation, 32–33, 52–53, 54–55, 128, 208–9; ef-
    forts to model engineering as, xviii, 32–33, 101,
    173; lack of stimulus to radical innovation, 34,
    78, 127–28, 220–22; radio as applied science,
    154, 243n6
"scientific management," 101. *See* Taylor,
    Frederick W.
Scientific American, 124
Selden, George, 95–96, 126
Sherman Antitrust Act (1890), 63, 149
Silliman, Benjamin Jr., 58
Silzer, George, 162, 163
Sinclair, H. H., 182, 185
Skelton, Owen, 201
Sloan, Alfred P. Jr., 79, 97; General Motors, reor-
    ganization by, 99–100; innovation, cautious ap-
    proach to, 100–101
Smeaton, John, 114; Smeaton coefficient, 114,
    115, 116, 240n20
Smith, Al, 163
Smithsonian Institution, 106, 108; Henry's advice
    to Bell, 43; Langley's aviation research, 106–7,
    120; dispute with Wright brothers, 11, 127
Standard Oil Company, 6, 57; breakup (1911),
    63, 64; consolidation of, 61; early innovation
    in, 63–65; opposition to Burton process, 11, 69;
    reorganizations of, 63, 69, 232n11. *See also*
    Standard Oil of Indiana